University of Plymouth
Charles Seale Hayne Library
Subject to status this item may be renewed
via your Primo account

http:/primo.plymouth.ac.uk
Tel: (01752) 588588

AGRICULTURE AND
EU ENVIRONMENTAL LAW

Agriculture and
EU Environmental Law

BRIAN JACK
Queens University Belfast, Belfast

ASHGATE

Published by
Ashgate Publishing Limited
Wey Court East
Union Road
Farnham
Surrey, GU9 7PT
England

Ashgate Publishing Company
110 Cherry Street
Suite 3-1
Burlington
VT 05401-3818
USA

www.ashgate.com

British Library Cataloguing in Publication Data
Jack, Brian.
  Agriculture and EU environmental law.
  1. Agriculture and state--Environmental aspects--European
  Union countries. 2. Agriculture--Environmental aspects--
  European Union countries. 3. Environmental policy--
  European Union countries.
  I. Title
  338.1'81'094-dc22

Library of Congress Cataloging-in-Publication Data
Jack, Brian.
  Agriculture and EU environmental law / by Brian Jack.
      p. cm.
  Includes index.
  ISBN 978-0-7546-4540-5 (hardback) -- ISBN 978-0-7546-9241-6 (ebook)
  1. Agricultural laws and legislation--European Union countries. 2. Environmental law--
European Union countries. I. Title. II. Title: Agriculture and European Union environmental law.
  KJE6605.J33 2009
  343.24'076--dc22

                                                                                    2009026739

Transfered to Digital Printing in 2012

ISBN 978-0-7546-4540-5

Printed in the United Kingdom by Henry Ling Limited, at the Dorset Press, Dorchester, DT1 1HD

Contents

List of Tables

Table of Cases

EUROPEAN COURT OF JUSTICE

COURT OF FIRST INSTANCE

UNITED KINGDOM

WORLD TRADE ORGANISATION

Table of Legislation

EUROPEAN UNION

EC Treaty Articles

Regulations

Directives

INTERNATIONAL

UNITED KINGDOM

Acts of the United Kingdom Parliament

Acts of the Scottish Parliament

Northern Ireland Primary Legislation

Statutory Instruments

Preface

Agricultural law and policy in the United Kingdom has been strongly influenced by the European Union's Common Agricultural Policy (the 'CAP') ever since the United Kingdom joined the, then, European Economic Community in 1973. At European level the CAP also occupies a central position within the European Union. It is a policy that has undergone significant change in recent years. When originally established, in the early 1960s, its principal goals were to encourage increased agricultural productivity and to protect farm incomes. In recent years a number of reform packages have been implemented that have gradually changed the CAP's emphasis from quantity to quality and from produce support to producer support. The most recent reform package was the December 2008 CAP Health Check. The various reform packages have been triggered by a range of issues, both budgetary and trade related. Additionally, increasing awareness of the environmental side-effects associated with intensive agriculture and of the need to provide greater support for farming methods protecting and enhancing the environment have also been important factors. Indeed, as readers will be well aware, environmental issues generally have climbed the political agenda and become much more important to politicians and policy makers. At European level the 1986 Single European Act amended the European Community Treaty to require that environmental protection requirements should be a component of the Community's other policies. Today, following subsequent amendments by the Treaty on European Union and Treaty of Amsterdam, Article 6 of the European Community Treaty provides the firmer instruction that 'environmental protection requirements must be integrated into the definition and implementation of Community policies and activities..., in particular with a view to promoting sustainable development'. Given agriculture's major role in the European countryside it should be no surprise that CAP is now viewed as having a major role in supporting many of the European Union's major environmental policy initiatives. The CAP Health Check, for example, acknowledges its central role in protecting and enhancing biodiversity, managing and protecting water resources and tackling climate change.

In practice CAP resembles a super tanker, which requires a great deal of time and effort to change direction. This book critically examines the directional change that has occurred in response to the European Community Treaty's requirement that environmental protection requirements be implemented into the definition and implementation of the CAP and provides critical analysis of the steps that have been taken. The book commences with two background chapters. Chapter 1 explains the development and operation of the CAP and of its financial mechanisms. In contrast, Chapter 2 analyses the environmental implications of the initial policy goals promoting greater intensification and increased productivity. This chapter also explores the developing awareness of environmental issues amongst European policy makers. In practice, the CAP has been built upon two, albeit uneven, pillars. The first pillar is principally concerned with production policy. The second pillar initially concentrated upon agricultural structures, encouraging the development of larger, more modern, farms that would be more profitable for their owners. More recently this structural policy has been broadened into a wider measure promoting rural development. The measures adopted to integrate environmental protection issues within each of these pillars are analysed in the subsequent chapters. Chapter 3 examines CAP's agricultural production policies, analysing the influence of measures designed to discourage intensification and the role played by cross compliance and set aside requirements. Chapters 4 to 6 explore the European Union's rural development policy. They examine individual measures and consider their effectiveness in promoting the introduction, or continuation, of agricultural methods protecting or enhancing the countryside. Chapter 4 provides an introduction to the European Union's rural development policy and examines the Less Favoured Areas scheme. Chapter five concentrates upon the European Union's agri-environment scheme, which seeks to promote long term environmental management on farms across the European Union. Chapter 6 explores the contribution made by the Organic Farming scheme. Like all other economic sectors agriculture must also respect the European Union's general environmental laws. With that in mind Chapters 7 and 8 examine specific directives that are particularly relevant to agriculture. Chapter 7 concentrates upon nature conservation, where the 1979 Wild Birds Directive and 1992 Habitats Directive have a central role, whilst Chapter 8 considers a range of other directives that regulate agricultural pollution. Chapter 8 examines the issue of food safety. Article 174 of the European Community Treaty recognises the protection of public health as being a central objective of the Community environmental policy, whilst equally, under Article 152 of that Treaty, this should be a core objective of all Community policies. On the topic of food safety the European Union has adopted an integrated, farm to table, approach, emphasising the important initial role of the farmer. Finally the book concludes in Chapter 10 with the issue of international trade. As the chapter shows, international trade agreements have had a major influence on the manner in which the CAP has been reformed in recent years and are likely to continue to do so in the future. The chapter also explores the role played by environmental considerations within international trade agreements and the measures in place to prevent animals or plants carrying infectious diseases from being imported into the European Union.

The time period covered within the book runs from the creation of the European Economic Community in 1957, to the creation of the European Union in 2004 by the Treaty on European Union. As a result of that Treaty the European Economic Community simply became known as the European Community. As many readers will be aware the Treaty on European Union provided for the continuation of the European Community within the broader structure of the European Union. To avoid causing confusion subsequent chapters will simply use the terms European

Community or the Community. Many readers will also be aware that one of the changes introduced by the Treaty of Amsterdam in 1999 was the renumbering of the articles of the European Community Treaty. It is these new article numbers that have been used throughout this book.

Lastly, I would like to thank friends and colleagues in the Law School for their on-going support during my work upon this book. I would also like to thank Alison Kirk, Nikki Selmes, Norman Gilligan and Sarah Horsley at Ashgate Publishing for all their help. More broadly I would like to thank Kluwer Law International for allowing me to use my article 'Tackling Eutrophication: The Implications of a Precautionary Approach' (published in [2006] 15 *European Environmental Law Review*) as a base from which to develop my discussions on nitrate pollution in Chapter 8. Thanks are also due to the European Commission and to the European Environment Agency for granting me permission to reproduce the tables printed in Chapters 5 and 7. Finally, and most importantly of all, a huge thank you is due to my family for all your support and encouragement throughout this project.

I have endeavoured to state the law as in May 2009, thought I have also been able to incorporate some later developments.

Chapter 1
Establishing the Common Agricultural Policy

Agriculture in the EC Treaty

The Common Agricultural Policy has been a major influence upon European agriculture for more than forty years. The reasons for its existence can be traced back to the creation of the European Economic Community, as it then was, in 1957. Some have suggested that it resulted from a compromise between France and Germany, that France would only allow German industrial exports to have free access to its markets if French agricultural produce gained similar access to German markets.[1] For manufactured goods, the European Community Treaty ('the EC Treaty') sought to achieve free movement of goods by removing import tariffs and quotas on trade between Member States. However, this approach alone was unlikely to be successful for agricultural produce, since Member State governments had a long history of intervening to protect national agricultural industries. As one commentator has pointed out, merely lifting quotas and tariffs, without also tackling the impact of these various protective measures, would not have resulted in the free circulation of agricultural produce.[2] Instead, common measures were required, to ensure that producers received equal support and protection in each Member State.

Prior to the establishment of the European Community, France actively supported the idea of establishing a 'European Agricultural Community,' which would have placed European agricultural production under the control of a supranational European institution, the High Authority. Negotiations concerning the proposed European Agricultural Community, the so-called 'Green pool', were conducted between 1952 and 1954, but no agreement was achieved on agricultural integration.[3] In view of this lack of agreement, it is perhaps no surprise that when

---

1   Grant, W. (1997), *The Common Agricultural Policy* (London: Macmillan) 63. For an alternative view see Ackrill, R. (2000), *The Common Agricultural Policy* (Sheffield: Sheffield University Press) 39.

2   Fennell, R. (1997), *The Common Agricultural Policy: Continuity and Change* (Oxford: Clarendon Press) 13.

3   See Fearne, A. (1997), The History and Development of the CAP 1945–1990, in Ritson, C. and Harvey D.R., (eds) *The Common Agricultural Policy*, 2nd ed. (Wallingford, Oxon: CAB International) 12 or Tracey, M. (1982). *Government and Agriculture in Western Europe 1880–1988* (Brighton: Harvester Wheatsheaf) 243.

the EC Treaty was signed in March 1957, it only dealt with agriculture in vague terms. The EC Treaty confirmed that the six original Member States wished to introduce a common policy on agriculture, but did not try to set out the terms of that policy. As one commentator has stated, the Treaty was, basically, an 'agreement to agree' on agricultural matters.[4] Instead, Article 37 of the EC Treaty still provides that, immediately upon the Treaty coming into force, the European Commission should convene a conference of the Member States in order that their agricultural policies, their resources and their needs could be assessed. The Commission held this conference at Stressa, Italy, in July 1958 and in so doing began to lay the foundations for the development of the Common Agricultural Policy.[5]

Nevertheless, the EC Treaty did actually set out a framework, within which the future policy would be developed. Article 33 EC specifies that the objectives of the policy are:

a. To increase agricultural productivity by promoting technical progress and ensuring the rational development of agricultural production and the optimum utilisation of the factors of production, in particular labour
b. Thus to ensure a fair standard of living for the agricultural community, in particular by increasing the individual earnings of persons engaged in agriculture
c. To stabilise markets
d. To assure the availability of supplies
e. To ensure that supplies reach consumers at reasonable prices

These objectives reflected the principal factors that had motivated the Member States own national policies prior to the establishment of a common policy.[6] They promoted economic growth amongst the Member States, both individually and collectively, whilst also reflecting the need to protect farm incomes and to ensure that adequate food supplies were available at reasonable prices.

Each individual objective is given equal weighting within Article 33, so none is given any greater importance. However, it can also be observed that the objectives are somewhat contradictory in nature. This was acknowledged by the European Court of Justice in 1968, when it noted that 'these objectives, which are

---

4   Ritson, C. (1971), *Agricultural Policy and the Common Market* (London: Chatham House Press) 104.

5   For discussion of the Stressa conference, see: Neville-Rolfe, E. (1984), *The Politics of Agriculture in the European Community* (London: Policy Studies Institute) 195, Tracey, M. (1982), *Agriculture in Western Europe* (St Albans: Granada Publishing) 269, Fennell, R. *The Common Agricultural Policy: Continuity and Change*, footnote 2 above, 17 or Fearne, R. The History and Development of the CAP 1945–1990 in Ritson, C., and Harvey, D.R. (eds) *The Common Agricultural Policy*, at footnote 3 above, 16.

6   McMahon, J.A. (1999), *Law of the Common Agricultural Policy* (Harlow: Pearson Education Ltd) 24.

intended to safeguard the interests of both farmers and consumers, may not all be simultaneously and fully attained.'[7] The Court, subsequently, set out the following formula for balancing the competing demands of these objectives:[8]

> In pursuing these objectives, the Community Institutions must secure the permanent harmonisation made necessary by any conflicts between these aims taken individually and, where necessary, allow any one of them temporary priority in order to satisfy the demands of the economic factors or conditions in view of which their decisions are made.

However, as several commentators have pointed out, the Court has generally given priority to the protection of farm incomes and has elevated the interests of producers over those of consumers.[9] This is despite the fact that Article 34(3) EC provides that European agricultural policy 'shall exclude any discrimination between producers or consumers within the Community'. The Court has interpreted this article as meaning that there should be no discrimination between individual producers or between individual consumers, rather than with situations in which Community measures unduly favour producers as a class over consumers as a class.[10]

The objectives set out in Article 33 EC provide a yardstick against which individual legislative measures might be measured. However, they also provide Community institutions with broad discretion in developing the individual legislative measures that contributed to the creation of a common policy on agriculture. Traditionally, the Court has been reluctant to entertain judicial reviews of legislation where the Community institutions have exercised such broad discretion. In such cases, the Court has previously noted that it would merely consider whether there had been a patent error or misuse of power.[11]

Subsequent chapters highlight the fact that the Common Agricultural Policy has, in recent years, undergone a number of significant reforms. As a result, it now incorporates a much broader range of policy goals than was the case in the original policy. For example, environmental protection is an important issue within the current Common Agricultural Policy. However, it was not an important influence upon policy makers in the 1950s. Similar comments could equally be made about

---

7   Case 5/67 *Beus v Hauptzollamt München* [1968] ECR 83 at 98.

8   Case 5/73 *Balkan-Import-Export GmbH v Hauptzollamt Berlin-Packhof* [1973] ECR 1091 at 1112.

9   Snyder, F.G. (1985), *Law of the Common Agricultural Policy* (London: Sweet and Maxwell Ltd) 20 and Usher, J.A. (2001), *EC Agricultural Law*, 2nd ed. (Oxford: Oxford University Press) 34.

10   See Case 5/73 footnote 8 above at 1112.

11   See Case 57/72 *Westzucker GmbH v Einfuhr und Vorratsstelle für Zucker* [1973] ECR 321 at 340 and Case 78/74 *Deuka, Deutsche Kraftfutter GmbH and B.J. Stolp v Einfuhr und Vorratsstelle für Getreide und Futtermittel* [1975] ECR 421 at 432.

other issues such as animal welfare or food quality. Consequently, it has been argued that the objectives set out in Article 33 EC no longer fully reflect the true goals of the modern Common Agricultural Policy.[12]

Defining the Scope of Agricultural Production

Article 32(1) EC states that the common market should be extended to include agriculture and trade in agricultural products. 'Agriculture' is not defined within the EC Treaty. Instead, Article 32(1) EC defines 'agricultural products' as meaning 'products of the soil, of stock farming and of fisheries and products of first stage processing directly related to these products'. In Case 185/73 *Hauptzollamt Bielefield v Konig*, the European Court of Justice concluded that, in order for produce to be regarded as being the product of first stage processing, there had to be 'a clear economic interdependence between basic products and products resulting from the a productive process'.[13] In other words, agricultural produce processed within a production process will no longer be regarded as being an agricultural product if the cost of the original, agricultural raw material is marginal to the cost of processing.[14]

The practical impact of the definition of 'agricultural products' in the EC Treaty is that the scope of the Common Agricultural Policy is not limited to the raw materials produced by farmers, but also includes activities such as the manufacture of dairy products, the slaughtering, preserving and preparing of meat and grain milling.[15]

Article 32(3) EC provides that a list of the agricultural products that come within the scope of the Common Agricultural Policy is set out in Annex I of the Treaty. The article also empowered the Council to add to this list within two years of the Treaty's entry into force.[16] This list today provides for most agricultural products to be regulated by the CAP. The principal exceptions are ethyl alcohol, honey, potatoes, wood and wool.[17]

---

12   See, for example, J.A. McMahon, The Common Agricultural Policy: From Quantity to Quality. [2002] 53 *Northern Ireland Legal Quarterly* 9.

13   [1974] ECR 607 at 618.

14   Snyder F.G., footnote 9 above, 17.

15   Snyder F.G. (1990), *New Directions in European Community Law* (London: George Weidenfield & Nicholson Ltd) 105.

16   This it did in Council Regulation 7a, [1961] JO 71, OJ Special Edition 1959–1962, 68. For an examination of the legality of this regulation see Case 185/73 *Hauptzollamt Bielefeld v König* [1974] ECR 607.

17   Although the Common Agricultural Policy does regulate sheepmeat, it does not cover wool.

The Common Organisation of Agricultural Markets

Article 34(2) EC provides for the establishment of a common organisation of agricultural markets for each of the agricultural products covered by the Common Agricultural Policy. The article specifies that, depending upon the product concerned, this could be achieved by adopting one of three formats. These are:

a. The adoption of common rules on competition
b. Or the compulsory co-ordination of the various national market organisations
c. Or the creation of a European market organisation

The favoured approach has been the establishment of European market organisations. These have been known as 'common organisations of the market' or 'common organisations'. Today, some 22 such common organisations exist, each governing different agricultural commodities.[18] The role of each common organisation is explained by Article 34(3) EC, which provides that they should:

> ...include all measures required to attain the objectives set out in Article [33], in particular regulation of prices, aids for the production and marketing of various products, storage and carrying over arrangements and common machinery for stabilising imports and exports.

Each common organisation was established by a Council regulation, which set out the measures that apply to the produce governed by that common organisation. In every case these measures were initially constructed around three core principles.[19] Firstly, that a single market would be created, so that there would be a system of common prices across the Community and both barriers to trade and distortions in competition between farmers in different Member States would be eliminated.[20] Secondly, that a principle of Community preference was created, to ensure that Community produce would be favoured over imported produce. Thirdly, that the common organisations should operate on the basis of financial solidarity, so that all Member States would contribute to the cost of operating the common organisations.

---

18   Barents, R. (1994), *The Agricultural Law of the EC* (Dordrecht, the Netherlands: Kluwer Law and Taxation Publishers) 74.

19   See European Commission (1982), *The Agricultural Policy of the European Community*, 3rd ed. (Brussels: European Commission) 14.

20   For discussion of the problems of operating common prices in the absence of a common currency see Ritson, C. and Swinbank, A. (1997), Europe's Green Money, in Ritson, C. and Harvey, D.R., *The Common Agricultural Policy*, 2nd ed. (Wallingford, Oxon: CAB International) 115.

Many of the common organisations have been radically reformed in recent years.[21] They were all, initially, based upon the market principle, that farm incomes should be determined by market prices.[22] Equally, however, these market prices were themselves often manipulated by the common organisations.[23] One commentator pointed out that the initial common organisations could generally be divided into three groups, depending upon the level of price support that they provided for farmers.[24] These groups distinguished between common organisations that provided farmers with complete price guarantees, those that provided partial price guarantees and those that provided no price guarantees.

Major commodity groups such as cereals, dairy produce and beef fell into first category, with their producers being accorded significant price and market guarantees. Cereal crops were of particular importance since they provided feedstuffs for other sectors of agriculture, such as poultry and pig meat. Consequently, cereal prices also influenced commodity prices and farm incomes within these sectors. The common organisation in cereals was established around three prices: a target price, an intervention price and a threshold price.[25] Each of these prices was reviewed annually by the Council of Ministers. This price structure was also applied in the other common organisations which provided full price guarantees to farmers.

The target price was the maximum price that, it was hoped, produce would attain upon Community markets, but beyond which those prices could no longer be considered reasonable for consumers. In contrast, the intervention price was a minimum price, below which it was considered producers would be unable to earn reasonable incomes. Member States were required to designate national intervention agencies that would purchase cereals when the market price fell to this level.[26] In this way the intervention price provided a floor for the market price and the intervention system also provided a market for surplus produce. Finally, there was the threshold price applied at the Community's borders – generally fixed at the same level as the target price. It was used to determine the price of produce imported into the Community. Variable levels were attached to the actual price of the, invariably, cheaper imported produce to bring those prices up to the

---

21   See Chapter 3 for a discussion of the process of reform.

22   Melchoir, M. (1983), The Common Agricultural Policy, in European Commission, *Thirty Years of Community Law* (Luxembourg: The Office for Official Publications of the European Communities) 73.

23   For a detailed analysis of the operation of the common organisations within the initial Common Agricultural Policy see either Snyder, F.G., footnote 9 chapter 4 or Usher, J.A. (1988), *Legal Aspects of Agriculture in the European Community* (Oxford: Clarendon Press) chapter 4.

24   Snyder, F., footnote 9 above, 73.

25   Established initially by Council Regulation 120/67 [1967] JO L22/69.

26   See Council Regulation 729/70, [1970] JO L94/13, Article 4. In the United Kingdom the Intervention Board for Agricultural Produce was established for this purpose under Article 6 of the European Communities Act 1972.

threshold price. The actual market price in the Community usually floated between the target and intervention prices set by the Community. By linking the threshold price to the target price and raising import prices to this level, the Community prevented imported produce from undercutting Community prices. Additionally, the threshold price also acted as a brake upon market prices in the Community. If those prices began to exceed the Community's target price, imported produce would then become cheaper than Community produce and this, in turn, could be expected to lower the Community's market prices. Allied to this price system, the Community also established a system of export refunds. These refunds sought to encourage farmers to export surplus produce to other markets. They ensured that farmers were compensated where the export price that they obtained was lower than the prevailing Community price.

Collectively, this price structure contributed to the three core principles of the initial policy. It sought to establish a system of common prices within an agricultural single market. The threshold price and the use of variable import levels discouraged imports, thereby supporting the principle of Community preference. The Community itself refunded Member States expenditure upon intervention payments and export refunds – thus ensuring a system of financial solidarity through the Community budget.

The price structure also enabled the Community to address a number of the problems that were being experienced in European agriculture. Agriculture invariably suffers from a low income elasticity of demand.[27] Although incomes may increase in the general population this will not necessarily result in more food being purchased, since there is a limit to the amount of food people can consume. Instead, consumers would increasingly switch to more expensive, higher quality, food or to more convenient, ready to eat, food. However, in the early 1960s, food was generally plentiful and agricultural production was increasing more quickly than population growth. This posed the risk that plentiful supplies would deflate agricultural prices and cause farm incomes to fall.

Equally, farm incomes were already causing concern to national governments in the late 1950s and 1960s. At a time of general economic expansion, farm incomes had not kept pace with the incomes available in other sectors of the economy. Unless this problem was addressed these governments faced the prospect of having to deal with the social problems that would accompany a large movement of people from farming into urban areas to seek employment in higher income jobs.[28]

---

27   See further Marsh, J.S., Ritson, C. (1971), *Agricultural Policy and the Common Market* (London: Chatham House Press) 15.

28   See European Economic Community Commission (1960), *Proposals For The Working Out and Putting into Effect of the Common Agricultural Policy in Application of Article 43 of the Treaty Establishing the European Economic Community* (Brussels: European Economic Community Commission) chapter 1.

Finally, the Member States were already beginning to face the issue of surplus production. It has been estimated that agricultural production in 1960 already met 87 per cent of the food requirements of those Member States.[29] This created a dilemma. As a result of protectionist policies, agricultural prices in the Member States were generally above world price levels.[30] If the Community wished to maintain such price levels it would also need to create a mechanism which would enable it to export its surpluses and to compete in foreign markets. The alternative, of limiting support to the amount of produce that was actually sold within the Community, would again reduce prices and farm incomes.[31]

The introduction of the price structure enabled the Common Agricultural Policy to address such issues. The intervention system ensured that farmers were guaranteed a minimum price for their produce and also a market for surplus produce. Equally, the threshold price and the use of variable import levies prevented imported produce from undercutting Community prices and reducing the market for Community produce, whilst the use of export refunds enabled producers to export surplus produce without loss of income.

Although this price structure provided significant income guarantees for farmers who produced commodities such as cereals, dairy produce and beef, it should also be recalled that not all common organisations provided this level of protection. For example, common organisations in poultry and eggs provided no system of internal price support.[32] Instead, they sought to provide protection by adding a levy to imported products which took account of the difference in the price of the amount of grain that would be required to produce a given amount of poultry meat or eggs in the Community and in the country of origin.[33]

Creating a Structural Policy for Agriculture

The EC regulations establishing the common organisations collectively set out the Community's agricultural production policies. The protection of farm incomes was an important factor behind these policies. However, agricultural production policies alone could not solve the problem of low farm incomes. Even if agricultural prices were set at levels needed to maintain small farmers, it would be larger, more efficient, farmers that would be the principal beneficiaries. The Community, therefore, put in place a separate structural policy to help with this

---

29   *Ibid.* 10.

30   See Fennell, R., footnote 2 above, 31.

31   United Nations Economic Commission for Europe (1961), *Economic Survey of Europe in 1960* Geneva: United Nations Economic Commission for Europe) 27.

32   These common organisations were initially created in 1967 by Council Regulation 122/67, [1967] JO 2293 in relation to eggs and Council Regulation 123/67, [1967] JO 17 for poultry meat.

33   See Snyder, F., footnote 9 above, 87.

issue. This structural policy became known as the 'second pillar' of the Common Agricultural Policy.

The EC Treaty itself recognises the need for a structural policy. Article 34(2) EC provides that, in developing a common agricultural policy, account should be taken of:

> the particular nature of agricultural activity, which results from the social structure of agriculture and from structural and natural disparities between the various agricultural regions.

In practice, farm incomes varied widely throughout the Community due to factors such as the quality of land and the skill of the farmer. Another important factor was the size of individual farms. Larger farms not only offered the possibility of greater productivity, but also of more efficient use of labour, capital and machinery. Farm incomes on larger farms were actually comparable to those available in other sectors of the economy.[34] However, in 1960, more than two thirds of farms in the six original Member States were less than ten acres in size and completely unable to produce such incomes for their owners.[35]

In its initial proposals for a Common Agricultural Policy, the Commission identified the need to enlarge farms that were too small to productively employ families and to provide an adequate income.[36] Similarly, it pointed to a need to regroup fragmented holdings.[37] The Commission feared that without such structural changes future economic growth would be limited to industrial areas and areas that already had more advanced agricultural structures. It, therefore, sought to promote a structural policy that would create larger farms capable of embracing new technologies and of becoming more productive, thus producing higher incomes for their farmers.[38]

Prior to the creation of a Community structural policy, most Member States had operated their own structural policies. These national policies affected agricultural output in different ways, something that threatened to distort competition within the Common Agricultural Policy.[39] Initially, the Community sought to co-ordinate the various national policies. Under Regulation 17/64,[40] applications for structural aid were submitted to the Commission by each Member State and required the

---

34 See Milward, A.S. (1995), *The European Rescue of the Nation State* (London: Routledge) 225, or Bowler, I.R. (1985), *Agriculture under the Common Agricultural Policy, A Geography* (Manchester: Manchester University Press) 204.

35 European Economic Community Commission, footnote 28 above, 15.

36 *Ibid.* 35.

37 *Ibid.*

38 *Ibid.* chapter 3.

39 Hill, B. (1984), *The Common Agricultural Policy: Past, Present and Future* (London: Methuen & Co. Ltd) 27.

40 Council Regulation 17/64 [1964] JO 586/103.

prior approval of that Member State. Projects were required to contribute to the 'improvement of the social and economic conditions of persons engaged in agriculture.'[41] Priority was given to projects which were 'part of a comprehensive system of measures aimed at encouraging the harmonious development of the overall economy of the region where such projects will be carried out.'[42]

The Community, however, was keen to develop its own structural policy measures. In a sector dominated by small, inefficient, farms one solution would have been to promote policies that encouraged small farmers to leave agriculture for other careers. Their farms could then be amalgamated with others farms to create larger, more efficient farms. This was, indeed, largely what was initially suggested. In 1968, the European Commission proposed reforms that sought to reduce the number of people working in agriculture by five million between 1970 and 1980.[43] Additionally, in that time, agricultural land would be reduced by five million hectares, in order to prevent future agricultural surpluses.[44] It has been pointed out that these proposals were largely founded upon economic aims.[45] On the one hand, production would be halted on less productive lands, whilst, on the other hand, improved farming structures and greater use of modern production methods would be encouraged in areas with better quality lands. The objective being to ensure that a much smaller agricultural sector would enjoy better incomes.[46] However, the proposals led to large scale public protests and were never adopted.[47]

Instead, the Community's subsequent structural policies stemmed from a number of directives adopted in the 1970s. These directives recognised that the causes, nature and gravity of structural problems, varied enormously from region to region. They sought to provide a common framework within which Member States would have a measure of discretion to decide which were most suitable to local conditions.[48] In 1972, the Community adopted three structural directives. Directive 72/159 authorised Member States to provide financial assistance for investments upon farms which were suitable for development.[49] Directive 72/160 authorised Member States to make early retirement payments to farmers and farm workers who were aged between 55 and national retirement age.[50] This Directive sought to encourage these farmers to retire so that their land could either be amalgamated

---

41   *Ibid.* Article 14(2).
42   *Ibid.* Article 15(1).
43   Commission of the European Communities (1968), *Memorandum on the Reform of Agriculture in the European Economic Community*, COM (68) 1000.
44   *Ibid.*
45   Cardwell, M. (2004), *The European Model of Agriculture* (Oxford: Oxford University Press) 26.
46   *Ibid.*
47   See Neville-Rolfe, E., footnote 5 above, 227 or Snyder, F., footnote 9 above, 155.
48   Snyder, F., footnote 9 above, 160.
49   Council Directive 72/159 [1972] OJ L96/1.
50   Council Directive 72/160 [1972] OJ L96/9.

with neighbouring farms, thus creating more economically viable farm units, or used for non-agricultural purposes. Thirdly, Directive 72/161 authorised Member States to provide socio-economic education and training for farmers.[51] This was designed to help farmers to take decisions about whether or not to remain in farming and would also provide training to those who did remain.

Later, in 1975, the Council also adopted Directive 75/268 on mountain and hill farming and farming in less favoured areas.[52] This Directive was a direct result of the United Kingdom's accession to the Community. The United Kingdom had had a long history of providing financial support to hill farmers. For example, the 1946 Hill Farming Act had continued a policy of providing subsidies to these farmers which the United Kingdom had initially introduced during the Second World War.[53] The adoption of Directive 75/268 enabled these policies to be continued, as part of the CAP. The Directive sought to maintain farming and rural communities in disadvantaged rural areas and to prevent damage to the countryside caused by the abandonment of agricultural land. Member States were required to identify the areas of disadvantaged agricultural land within which the Directive would apply. They were then authorised to make direct income payments to farmers located within these areas and to provide financial grants for farm development works or for diversification into tourist or craft industry programmes.

Directive 75/268 enabled the Community to make particular provision for farmers located in regions that had unfavourable agricultural conditions. In the late 1970s and early 1980s, the Community adopted further measures to support agricultural development in specific geographical regions. This began with a package of measures, collectively known as 'the Mediterranean Package', that were designed to improve agricultural structures in Italy and Southern France. These measures included Directive 78/627, which established a programme to encourage the restructuring and conversion of vineyards in Mediterranean areas of France;[54] Regulation 270/79 to encourage the development of farm advisory services in Italy[55]; Regulation 1944/81, which established a common measure for the adoption and modernisation of the structure for the production of beef, veal, sheep meat and goat meat in Italy.[56]

Community measures also targeted agricultural structures in other regions. For example, EC directives provided financial assistance for land-drainage projects in both Northern Ireland and the Republic of Ireland.[57] The Community

---

51   Council Directive 72/161 [1972] OJ L96/15.
52   Council Directive 75/268 [1975] OJ L128/1.
53   Marsden, T., Murdoch, J., Lowe, P., Munton, R., Flynn, A. (1993), *Constructing the Countryside* (London: University College Press) 58.
54   Council Directive 78/627 [1978] OJ L206/1.
55   Council Regulation 270/79 [1979] OJ L38/1.
56   Council Regulation 1944/81 [1981] OJ L38/6.
57   See, respectively, Council Directive 79/197 [1979] OJ L43/32 and Council Directive 78/628 [1978] OJ L206/5.

also introduced Regulations to encourage other reforms in agricultural structures within Northern Ireland and the western regions of the Republic of Ireland.[58] Equally, integrated development programmes were adopted for the Western Isles of Scotland,[59] the Department of Lozere in France[60] and agriculturally less favoured areas in Belgium.[61] These sought to promote agricultural development in these areas as well as supporting tourism, crafts and small industries. Each measure provided additional financial support for agricultural development within the regions identified for a limited period. This supplemented the support already available through the Community's structural policy for agriculture.

Moving from Structural Policy to Rural Development Policy

The Community replaced its 1972 agricultural structures Directives, along with the 1975 Less Favoured Area Directive, in 1985. It did so by adopting one regulation, Regulation 797/85 on improving the efficiency of agricultural structures.[62] This Regulation was designed to bring together most of the Community's structural policies in relation to agriculture within one regulation.[63] Regulation 797/85 made it compulsory for Member States to provide investment aid for farm improvements. Additionally, for the first time, the definition of farm improvements was extended to include projects designed to protect and improve the environment.[64] The Regulation also set out a range of other measures that Member States could, at their discretion, decide to adopt.

The centralising approach adopted in Regulation 797/85 remains in place today. However, the Community's structural policy has evolved into a more broadly based rural development policy. This evolution occurred in recognition of the fact that agriculture, today, is no longer the predominant employer in most rural areas. For example, in 1997, the European Commission reported that:[65]

> In terms of regional income and employment, agriculture no longer forms the main base of the rural economy. It represents only 5.5% of total employment on

---

58   See, respectively, Council Regulation 1942/81 [1981] OJ L197/17 and Council Regulation 1820/80 [1980] OJ L180/1.

59   Council Regulation 1939/81 [1981] OJ L197/6.

60   Council Regulation 1940/81 [1981] OJ L197/9.

61   Council Regulation 1941/81 [1981] OJ L197/13.

62   Council Regulation 797/85 [1985] OJ L93/1.

63   The early retirement scheme was a notable omission. This was, subsequently, rectified when the Community later adopted Regulation 1096/88 establishing a Community scheme to encourage the cessation of farming. [1988] OJ L110/1.

64   Council Regulation 797/85, Article 1.1.

65   European Commission, *Agenda 2000 Volume 1: For a Stronger and Wider Union*, COM (97) 2000 Final, 26.

average and only in very few regions is its share higher than 20%. The long term trend is a further drop in the number of farmers, at a rate of 2–3% per year.

Such recognition, of agriculture's weakened role within the rural economy, led to calls for the Community's agricultural structural policy to be widened into a vehicle that was more suited to the needs of the rural economy as a whole. For example, delegates to the first European Community conference on rural development, held in Cork in November 1996, agreed upon a ten-point declaration (the Cork Declaration) on future policy for rural areas. Point two of this declaration stated that:

> Rural Development policy must be multi-disciplinary in concept, and multi-sectoral in application, with a clear territorial dimension.... It must be based upon an integrated approach, encompassing within the same legal and policy framework: agricultural adjustment and development, economic diversification – notably small and medium scale industries and rural services – the management of natural resources, the enhancement of environmental functions, and the promotion of culture, tourism and recreation.

Against this background, it was no longer appropriate for Community funding for rural development to be channelled, principally, through structural policies aimed solely at farmers. Regulation 1257/99, therefore, replaced the Community's structural policy for agriculture with a more broadly based 'rural development plan'.[66] The Regulation set out a range of measures for which the Community would provide financial assistance. Member States were required to design rural development plans, covering a seven-year period from January 2000, to put these measures into operation.[67] In accordance with the principle of subsidiarity, Member States decided which measures were appropriate to their own particular circumstances. Initially, the only exception to this discretion was that all rural development plans had to include an agri-environment scheme, which made provision for farmers to enter contracts to provide environmental land management services.[68]

More recently, Regulation 1698/2005 has set in place a similar template for Rural Development policy.[69] It requires Member States to develop rural development

---

66   Council Regulation 1257/99 [1999] OJ L160/80.

67   *Ibid.* Articles 40–44. Article 41 provides 'rural development plans shall be drawn up at the geographical level deemed to be most appropriate.' In the United Kingdom separate rural development plans have been drawn up for England, Scotland, Wales and Northern Ireland.

68   *Ibid.* Article 43(2). See chapter 5 for a detailed examination of the agri-environment scheme.

69   Council Regulation 1698/2005 [2005] OJ L277/1.

plans for the period January 2008 to December 2013.[70] The Regulation provides a menu of 22 measures and again allows Member States to choose, within their rural development plans, those most suited to their rural areas. Agri-environment and animal welfare measures are exceptions. Member States continue to have an obligation to include measures to promote environmental land management and also must now include measures to support animal welfare within their rural development plans.

Regulation 1698/2005 has also sought to streamline the available rural development measures into four groups or 'Axes'. Axis 1 contains measures designed to boost the competitiveness of agriculture and forestry. Axis 2 sets out measures intended to support land management and to enhance the environment. In contrast, the measures contained in Axis 3 aim to support the social and economic fabric of rural areas as a whole. Similarly, Axis 4 seeks to assist collaborative projects to improve quality of life and economic prosperity in rural areas. The division of these measures into the Axes is more than a cosmetic exercise. The Axis structure is also important in regulating the amount of Community funding that can be provided. Regulation 1698/2005 requires that Axes 1 and 3 should each account for at least 10 per cent of all Community spending on rural development, whilst at least 25 per cent of that spending must be incurred upon measures contained within Axis 2.[71] A summary of the individual measures available within each Axis is provided in Table 1.1. It will be apparent from this table that, whilst policy measures do now make provision for broader rural issues, those targeted at agriculture continue to account for a substantial part of the Community's rural development policy.

Financing the Common Agricultural Policy: Initial Measures

Article 34(3) EC provides for one or more agricultural guidance and guarantee funds to be set up to finance the operation of the Common Agricultural Policy. Originally, the European Commission proposed that separate funds should be created for each common organisation.[72] However, in Council Regulation 25/62,[73] the Community, subsequently, elected to establish one fund, the European Agricultural Guidance and Guarantee Fund (EAGGF).[74]

The EAGGF became an integral part of the Community budget.[75] Income generated by the Common Agricultural Policy, through import tariffs and customs

---

70   *Ibid.* Article 6.

71   *Ibid.* Article 17.

72   Neville-Rolfe, E., footnote 5 above, 212.

73   OJ (sp. ed.) 1959–1962, 126.

74   The fund was also known by its French initials, FEOGA, representing Fonds Européen d'Orientation et de Garantie Agricole.

75   Council Regulation 729/70/EEC [1970] JO L94/13, Article 1(1).

Table 1.1   Summary of the Measures Contained within the Three Axes of the EC's Rural Development Policy for 2008–2013

Axis	Aid Measure
Axis 1: Improving Competitiveness	Vocational training and information The establishment of young farmers Early retirement payments to older farmers Use of advisory services (including for meeting standards) Setting up farm management, relief and advisory services Farm/forestry investment The development of agricultural processing and marketing Develop agricultural/forestry infrastructure Restoring agricultural production potential Temporary support in meeting agricultural standards Food quality incentive scheme Food quality promotion
Axis 2: Improving the environment and the countryside.	Less Favoured Area payments Payments to farmers managing important habitats (Natura 2000) Agri-environment land management payments Animal welfare payments Support for non-productive investments Support for afforestation Support to protect important habitats upon forested land (Natura 2000) Restoring forest production potential Support for non productive investments
Axis 3: Improving quality of life in rural areas and encouraging diversification of the economy.	Basic services for the rural economy and population (setting-up and infra-structure) Renovation and development of villages, protection and conservation of rural heritage Vocational training Capacity building for local development strategies Diversification into non-agricultural activities Support for micro-enterprises Encouragement of tourism activities Preservation and management of the natural heritage
Axis 4: Leader: Supporting locally based projects designed to improve quality of life and prosperity in rural areas.	The development and implementation of local development strategies within identified sub regional rural territories based upon public- private partnerships and bottom up approaches to decision making

This table is modelled upon one set out in European Commission (2004), *Fact Sheet: New Perspectives for EU Rural Development* (Luxembourg: Office for Official Publications of the European Communities) 12.

duties, formed part of the Community's own resources. These monies were collected by Member States and then forwarded to the Commission.[76] However, the Common Agricultural Policy was not self-financing, so the EAGGF also had access to the other budgetary resources of the Community.[77]

Regulation 17/64 provided for the EAGGF to be divided into a guarantee section and a guidance section.[78] Under the initial Common Agricultural Policy, the guarantee section was established to fund the Community's agricultural production policies, through the operation of the common organisations. In contrast, the guidance fund was intended to provide funding for structural policy measures.

The guarantee section financed the payment of export refunds and intervention payments. These payments were initially made to farmers by Member States' own national administrations. They in turn were then fully reimbursed by the EAGGF.[79] In contrast, the guidance section only provided for Member States to receive a partial refund of the expenditure they incurred on structural measures in agriculture. In the case of the Community's initial structural measures, only 25 per cent of Member States' expenditure on these measures was generally refunded. In contrast, the early-retirement measure enabled Member States to recover up to 60 per cent of their expenditure. The origins of this distinction, between a fully funded production policy and a partially funded structural policy, are not totally clear. Fennell has argued that some Member States had previously operated extensive national structural policies and were reluctant to assist in financing measures in others that had previously done little in this regard.[80] The Community's initial structural measures, however, have set an important precedent which remains in place today. The Community's rural development policy continues to operate on the basis that the Community will only provide partial funding for measures adopted within Member States.

Additionally, the funding arrangements for the Community's rural development policy have also enabled it to retain a regional aspect to its policies. Where, previously, the Community had adopted separate regulations to assist the development of agricultural structures in particular regions, its rural development policy, today, operates as a common framework across the whole Community. However, the Community does provide additional financial assistance to help particular regions fund rural development measures. This regional aspect to the rural development policy stems from Regulation 2052/88.[81] The Regulation

---

76   See for example, Council Decision 94/728/EC [1994] OJ L94/13, Article 8.

77   See Usher, J.A. (1988), *Legal Aspects of Agriculture in the European Community* (Oxford: Clarendon Press) 104.

78   Council Regulation 17/64 OJ (sp. ed.) 1963–1964, 103.

79   As long as the European Commission was satisfied that the payments had been made in accordance with the terms of the regulation governing the common organisation in question.

80   Fennell, R., footnote 2 above, 219.

81   Council Regulation 2052/88 [1988] OJ L185/9.

sought to rationalise all of the Community's structural policies, including the guidance section of the EAGGF.[82] It sought to ensure that the funds available under them would be allocated to areas of greatest need.[83] Regulation 2052/88 was subsequently replaced by Regulation 1260/99.[84] This Regulation divided the Community's regions into three categories:

Objective 1 Areas

Areas whose gross domestic product was less than 75 per cent of the Community average.[85] Effectively, these are regions whose economic development has lagged behind the rest of the Community.

Objective 2 Areas

Areas undergoing socio-economic change which do not qualify for Objective 1 status.[86] These include regions with declining rural areas. To qualify as Objective 2 areas such rural areas had to have a population density of less than 100 people per square kilometre or a percentage of agricultural employment that was equal to or higher than the Community average in any year since 1995. Additionally, to be recognised as being declining rural areas, these areas also had to have experienced average unemployment levels that were above the Community average or population decline in the period since 1985.

Objective 3 Areas

Areas not covered by Objective 1 in which funding may be provided to support the adoption and modernisation of policies and systems for education, training and employment.[87]

---

82   In addition to the EAGGF, the European Community operates several other structural funds which aim to encourage economic and social development and to reduce regional disparities. Examples of these include the European Social Fund, the European Regional Development Fund and the Cohesion Fund.

83   In addition to the Guidance section of the EAGGF other structural funds governed by the Regulation included the European Social Fund and the European Regional Development Fund.

84   Council Regulation 1260/99[1999] OJ L161/1.

85   *Ibid.* Article 3(1). Additionally, it also included some regions designated as objective one areas under Regulation 2052/88 that no longer met these criteria. Under Article 6(1) of the 1999 Regulation these now identified as being 'objective 1 regions in recovery' and continued to be eligible to receive objective one funding until 31st December 2005. Both Northern Ireland and the Highlands of Scotland came into this category.

86   *Ibid.* Article 4.

87   *Ibid.* Article 5.

Having divided the Community's regions into these categories, Regulation 1260/99 then provided for 69.7 per cent of all the Community's structural funding to be allocated to Objective 1 areas, with 11.5 per cent and 12.3 per cent, respectively, being ring-fenced for Objective 2 and Objective 3 areas.[88] Consequently, a greater share of the Community's rural development expenditure was been targeted towards Objective 1 areas. This influenced the level of Community financial support that was available to help finance rural development in each Member State. In regions recognised as having Objective 1 status, Regulation 1257/99 provided for the Community to contribute 50 per cent of the public expenditure cost of the rural developments measures adopted within those regions, and up to 25 per cent in other areas.[89] Additionally, the Community agreed to bear up to 75 per cent of public expenditure upon agri-environmental land management contracts within Objective 1 areas and up 50 per cent of public expenditure upon these contracts in other areas.[90] As noted in Chapter 5, the aim of the agri-environment scheme is to protect the environment and to improve the countryside.

Regulation 1257/99 also made a change to the financial structure of the Common Agricultural Policy. As noted above, the guarantee and guidance sections of the EAGGF had previously been given distinct functions – the guarantee section funding Community agricultural production policy and the guidance section funding structural policy measures. This clear division was ended by Regulation 1257/99, which gave responsibility for funding some aspects of rural development in areas without Objective 1 status to the guarantee section.[91] However, as noted below, Regulation 1290/2005 subsequently re-introduced this functional division.[92]

Financing the Modern Common Agricultural Policy

The changes introduced by Regulation 1257/99 formed part of a process of on-going reform of the CAP that will be examined in detail in Chapter 3. Indeed, they themselves have subsequently been reformed. Regulation 1290/2005, on the financing of the CAP, created a new financial framework for the CAP.[93] It created two new funds, the European Agricultural Guarantee Fund (EAGF) and the European Agricultural Fund for Rural Development (EAFRD).

The EAGF funds the Community's agricultural production policies, financing areas such as direct payments made to farmers, export refunds and intervention payments. The Community continues to fully fund eligible expenditure on these

---

88   *Ibid.* Article 7(2).
89   Council Regulation 1257/99, [1999] OJ L160/80, Article 47(2).
90   *Ibid.*
91   *Ibid.* Article 35.
92   Council Regulation 1290/2005 [2005] OJ L209/1.
93   *Ibid.*

measures. In contrast the EAFRD finances rural development policy measures. The operation of the EAFRD has been influenced by reforms to the Community's regional policy. Regulation 1083/2006 abolished the Objective 1 to 3 classifications that had previously determined funding for structural policy and regional policy measures.[94] In their place, the Community created three new regional policy objectives: Convergence, Regional Competitiveness and Employment, and Territorial Co-Operation.[95] The Convergence classification is similar to the previous Objective 1 category, in that it also seeks to promote economic development within the European Community's least developed regions. Once again, these are identified as being those with a Gross Domestic Product (GDP) per person below 75 per cent of the Community average. Additionally, it also covers 'phasing out' regions – those that would have been below this threshold had it been calculated on the basis of an EU of 15 Member States, as opposed the EU-25 Member States that existed when the Regulation came into force. In the United Kingdom, Cornwall, the Isles of Scilly, West Wales and the Welsh Valleys are all recognised as being convergence regions, whilst the Highlands and Islands of Scotland are 'phasing out' convergence regions.[96] In contrast, the Regional Competitiveness and Employment objective aims to encourage economic development and increased employment opportunities in regions that do not qualify as convergence regions. The European Territorial Co-Operation objective seeks to encourage cross-border and inter-regional co-operation. Collectively, these new objectives shape the direction of Community's structural regional policies for the period 2007 to 2013.

Today, the level of Community funding available for rural development policy measures continues to vary from region to region, depending upon each region's status within the Community's regional policy. The majority of funding is channelled into convergence regions. The Community reimburses up to 75 per cent of eligible public expenditure upon measures listed in Axis 1 (competitiveness) and Axis 3 (diversification and quality of life) when those measures are adopted in convergence regions.[97] Similarly, the Community reimburses up to 80 per cent of that expenditure upon measures listed in Axis 2 (improving the environment and the countryside) when those measures are adopted in convergence regions.[98] In contrast, the Community only reimburses up to 50 per cent and 55 per cent, respectively, of public expenditure upon similar measures in non-convergence areas.[99] Financial limits also apply to the total amount of Community funding that

---

94   Council Regulation 1083/2006 laying down provisions on the European Regional Development Fund, the European Social Fund and the Cohesion Fund, [2006] OJ L210/1.

95   *Ibid.* Articles 5–8.

96   See the European Commission, *Regional Policy Inforegio* [online], http://ec.europa.eu/regional_policy/atlas2007/uk/index_en.htm [last accessed 28 April 2008].

97   Council Regulation 1698/2005 on rural development, [2005] OJ L277/1, Article 70(3).

98   *Ibid.*

99   *Ibid.*

can be allocated to rural development measures in each Member State over the period 2007–2013.[100]

As it has from the outset, the Community continues to make only a partial contribution to the costs of operating its rural development policies. The practical impact of these policies has, therefore, been influenced by the degree to which Member States have been prepared to provide their own matched funding for these measures. In the past, from its inception until the late 1980s, the second pillar of the Common Agricultural Policy never accounted for more than 5 per cent of the Community's agricultural expenditure.[101] Today, the Community has begun to place greater financial emphasis upon its rural development policy, which now accounts for some 22 per cent of its total spending upon the Common Agricultural Policy.[102] One example of the increased emphasis upon rural development policy spending has been the introduction of compulsory 'modulation', whereby Member States have been required to reduce the amount of direct payments made to farmers by a fixed percentage, with the monies generated being used to fund rural development measures. Compulsory modulation was introduced in 2005, when a 3 per cent reduction was made on payments to farmers receiving more than €5,000 per annum in direct aid, increasing to 4 per cent in 2006 and to 5 per cent in 2007 and subsequent years.[103] In November 2008 the most recent agricultural reform package, the Health Check of the Common Agricultural Policy, agreed to increase the compulsory modulation rate to 7 per cent from 2009 and by a further one per cent a year to a ceiling of 10 per cent from 2012.[104] Additionally, under these reforms, a further 4 per cent reduction will also be applied from 2009 on farmers receiving €300,000, or more, in direct payments.[105] Member States can also elect to raise further monies for rural development, over the period 2007 to 2012, by deducting up to 20 per cent of direct payments under an additional voluntary modulation scheme.[106]

---

100   *Ibid.* Article 69(4) and Commission Decision 2007/383, [2007] OJ L142/21. A total of €88,294,374,687 in Community funds to be spent upon rural development measures over this period, of which €1,909,574,420 is available to the United Kingdom.

101   Grant, W., footnote 1 above, 71.

102   In 2006 European Community expenditure upon the Common Agricultural Policy was €54,226 million of which €11,931 million was spent upon Rural Development Policy. See European Commission (2008), *The Agricultural Situation in the European Union 2007 Report* [on-line], http://ec.europa.eu/agriculture/publi/agrep2007/index_en.htm [last accessed 18 February 2009] table 3.4.1.

103   Compulsory modulation was introduced by Council Regulation 1782/2003, [2003] OJ L270/1, Article 10.

104   Council Regulation 73/2009, [2009] OJ L30/16, Article 7(1).

105   *Ibid.* Article 7(2).

106   Under Council Regulation 1782/2003 laying down rules for voluntary modulation of direct payments, [2007] OJ L95/1.

Chapter 2

Agriculture and the European Environment

The Common Agricultural Policy and Agricultural Intensification

Agriculture is an important land user in Western Europe. It accounts for some 44 per cent of land use in the European Community, a figure that rises to over 70 per cent in Member States such as the United Kingdom and Ireland.[1] As such, agriculture has clearly had an important influence upon our rural environment. Historically, that influence was regarded as being positive, with farmers being considered to be 'the guardians of the countryside'. For example, the House of Lords Select Committee on the European Communities previously observed that: 'Farming created the countryside as it is today. Without farmers cultivating the soil and tending livestock, the countryside would gradually become a wilderness.'[2]

However, it is now widely recognised that modern agriculture has become an important source of environmental damage.[3] Within the European Community this environmental damage was encouraged both by the measures adopted within the Common Agricultural Policy ('the CAP') and by technological developments within the industry. As the European Environment Agency recently observed:[4]

> The common agricultural policy has been one of the important drivers of farm intensification and specialisation in the EU. Market pressures and technological development have also contributed to these trends which are very strong in some

---

1   European Commission (1999), *Agriculture, Environment, Rural Development: Facts and Figures – A Challenge for Agriculture* (Luxembourg: Office for Official Publications of the European Communities) 235. The figure of 44 per cent is based upon the EU-15, prior to the enlargement in both 2004 and 2007.

2   House of Lords Select Committee on the European Communities (1984), *20th Report: Agriculture and the Environment*, Session 1983–1984, (London: HMSO) paragraph 127.

3   See, generally, Shoard, M. (1980), *The Theft of the Countryside* (London: Maurice Temple Smith Ltd).

Body, R. (1982), *Agriculture – The Triumph and the Shame* (London: Maurice Temple Smith Ltd).

Rose, C. (1984), *Crisis and Conservation: Conflict in the British Countryside* (London: Penguin Books Ltd) Lowe, P. et al. (1986), *Countryside Conflicts: The Politics of Farming and Forestry Conservation* (Aldershot: Gower Publishing), Body, R. (1987), *Red or Green for Farmers* (Saffron Waldon: Broad Leys), Harvey, G. (1997), *The Killing of the Countryside* (London: Jonathon Cape).

4   European Environment Agency (2003), *Europe's Environment: The Third Assessment* (Copenhagen: European Environment Agency) 43.

sectors that benefit from little public support (e.g. pigs, poultry, and potatoes). Intensive farming has had significant impacts upon the environment.

As explained in Chapter 1, the Common Agricultural Policy initially provided important market guarantees for many farmers. Additionally, it established a system of relatively high commodity prices. Prior to the establishment of the Common Agricultural Policy these prices had varied quite considerably between the original Member States. In adopting common prices under the Common Agricultural Policy, the Member States chose to set agricultural commodity prices at a high level.[5] Consequently, when the United Kingdom and Ireland joined the Community, the Common Agricultural Policy ensured that farm incomes were determined by production levels and that farmers experienced a significant increase in commodity prices. Across the Community, the policy's market guarantees, together with high prices for agricultural commodities, made it more viable for farmers to adopt intensive and higher cost production methods.[6] It also encouraged them to make greater use of agricultural technology and to make maximum use of their land. Drainage, ploughing and reseeding could be conducted to bring previously unproductive land into production or to improve poorer land. Indeed, Community structural policy made financial assistance available for many such projects. The operation of the market mechanisms, together with high agricultural commodity prices, also led to increased land values throughout the Community.[7] This, in turn, created a strong incentive for farmers who had purchased or rented land to intensify their use of that land in order to recoup their outlay. In turn, this intensification also brought with it greater risk of environmental damage.

Agricultural Intensification and the Environment

Agricultural development, together with afforestation programmes and the expansion of urban areas has had a major impact upon rural landscapes. One report, in 1984, noted that in Great Britain during the period 1945 to 1984 these factors were associated with the loss of or significant damage to 30–50 per cent of

---

5   See further, Fennell, R. (1997), *The Common Agricultural Policy: Continuity and Change* (Oxford: Clarendon Press) 30.

6   See, for example, Harvey, D. (1991), *The CAP and Green Agriculture* (London: Institute for Public Policy Research).

7   See Baldock, D. (1984), *The CAP Price Policy and the Environment – An Explanatory Essay* (London: Institute for European Environmental Policy) 32, Harvey, D. footnote 6 above at 22 and Cheshire, P. *The Environmental Implications of European Agricultural Support Policies* in Baldock, D. and Conder, C. (eds) (1985), *Can the CAP Fit the Environment?* (London: Council for the Protection of Rural England) 12. For example, Harvey estimates that in 1991 the operation of the CAP support policies had inflated the value of agricultural land in the United Kingdom by an average of 46 per cent.

ancient woodlands, 80 per cent of unimproved grasslands, 40 per cent of lowland heath, 60 per cent of lowland raised bogs and 30 per cent of upland heaths and blanket bogs.[8]

The time period chosen by this report emphasises the fact that landscape change, and environmental damage, was already occurring under the national policies operating in Britain prior to our accession to the then European Economic Community. In more recent years this continued, and indeed escalated, under the Common Agricultural Policy. The following section will examine the environmental impact associated with a number of aspects of agricultural intensification.

Agricultural Specialisation

Forest is the natural vegetation cover of most of the British Isles. These forests began to be cleared in medieval times so that land could be cultivated and settled.[9] The cultivated areas became either grassland used for livestock grazing or arable land upon which crops were grown in rotation. The semi-natural grasslands that developed under this agricultural system were usually species rich environments that also supported a broad range of wild plants.[10] In turn, these wild plants supported a range of insect life.[11] Since the Second World War, however, agricultural policies have encouraged farmers to plough up semi-natural grasslands and replace them with rye grass based monocultures. This was because the botanical diversity of semi-natural grasslands limited their production potential, whereas the stronger growing rye grasses excluded slower growing wild plants and promised greater productivity. Consequently, however, these improved grasses contained fewer wild plants and were much less able to sustain wildlife such as insects. The Nature Conservancy Council, for example, has previously pointed out that whilst semi-natural pastures might typically support twenty species of butterfly, rye grass based improved grasses may well support none.[12] This, similarly, reduces their ability to support bird life. Aside from the limited availability of insect food sources, the nature of the grass sward will also impact upon its ability to provide a broader

---

8 Nature Conservancy Council (1984), *Nature Conservation in Great Britain* (Shrewsbury: Nature Conservancy Council) 49–59. See also Lowe, P. (1986), *Countryside Conflicts: The Politics of Farming, Forestry and Conservation* (Aldershot: Gower) 55.

9 See further, Green, B.H. Agricultural Intensification and the Loss of Habitat, Species and Amenity in British Grasslands: A Review of Historical Change and Assessment of Future Prospects. 1990 45 *Grass and Forage Science* 365.

10 *Ibid.* 368.

11 *Ibid.* See also Morris, M.G. Impacts of Agriculture on Southern Grasslands, in Jenkins, D. (ed.), (1984), *Agriculture and the Environment* (Cambridge: Institute for Terrestrial Ecology) 95.

12 Nature Conservancy Council (1977), *Nature Conservation and Agriculture* (London: Nature Conservancy Council) 12. See also Morris, M.G. footnote 11 above, 96 and Barr et al. (1993), *Countryside Survey 1990 Main Report* (London: Department of the Environment) 119.

habitat for birds. For example, redshank and snipe often feed in tufts of long grass near to their nests, but the improvement of grasslands limits the availability of such features.[13] Several studies have shown there to have been widespread loss of semi-natural grasslands in Britain in favour of improved grasses.[14] These improved grasslands have, also, often been associated with higher nitrogen fertiliser usage and with increased livestock numbers.[15] Biodiversity loss has also occurred as a result of the replacement of permanent grasslands by arable crops. In 1975, the European Community had 9 Member States. Between 1975 and 1995, the area of permanent grassland in those Member States decreased by 12 per cent, a total four million hectares.[16] The Common Agricultural Policy's production policies played an important part in influencing landowners to make this change. One example of this was the introduction of milk quotas in 1984, which led to a decrease in dairy cattle numbers and released land for arable farming.[17]

Hedgerow Loss

Hedgerow loss has also been an important. Hedges have been described as being, numerically, the most important farmland habitat for birds.[18] Studies have shown a clear correlation between bird numbers and hedge abundance.[19] However, increasing mechanisation on farms led to large-scale removal of farmland hedges as farmers strove to make more efficient use of agricultural machinery. The

---

13　O'Connor, R.J., and Shrubb, M. *Farming and Birds* (Cambridge: Cambridge University Press) 93.

14　See Nature Conservancy Council (1984), *Nature Conservation in Great Britain*, (Shrewsbury: Nature Conservancy Council). This report stated, at p.50, that 95 per cent of lowland grasslands in Great Britain lacked any significant wildlife interest and that only 3 per cent were undamaged by agricultural intensification. Similarly they refer, at p. 51, to significant damage or loss of 80 per cent of semi-natural lowland grasslands due to conversion to arable land or though the introduction of improved grasses. See also Fuller, R.M., The Changing Extent and Conservation Interest of Lowland Grasslands in England and Wales: A Review of Grassland Surveys 1930–1984, [1987] 40 *Biological Conservation* 281. He noted, at p. 297, that natural or semi-natural grasslands made up only 11 per cent of the lowland grassland area of England and Wales in 1987. Similarly Sinclair, G. (1983), *The Upland Landscape Survey* (Dyfed: Environment Information Services) estimated that a comparison of the periods 1872 to 1967 and 1967 to 1978 showed an eleven fold increase in the rate of removal of semi-natural vegetation.

15　Beaufoy, G., Baldock D., and Clark, J. (1994), *The Nature of Farming in Nine European Countries*, (London: Institute for European Environmental Policy) 47.

16　European Commission, (1999), *Agriculture, Environment, Rural Development: Facts and Figures – A Challenge for* Agriculture (Luxembourg: Office for Official Publications of the European Communities) 51.

17　*Ibid.*

18　Lack, P. (1992), *Birds on Lowland Farms* (London: HMSO) 32.

19　O'Connor, R.J. The Importance of Hedges to Songbirds, in Jenkins, D. (ed.) (1984), *Agriculture and the Environment*, (Cambridge: Institute of Terrestrial Ecology) 118.

Nature Conservancy Council previously estimated that between 1945 and 1974 some 140,000 miles of hedgerow were removed in Great Britain.[20] The highest levels of hedgerow loss were generally experienced in areas of arable farming. In contrast, in areas which supported livestock farming, hedgerows provided shelter for livestock. In Ireland, both Northern Ireland and the Republic of Ireland, agriculture is predominantly based upon livestock farming. Consequently, Ireland experienced a lesser degree of hedgerow loss than Great Britain.[21]

Where hedgerow loss has occurred, serious biodiversity loss would not occur if sufficient alternative habitats remained – such as rough grass, scrub and woodland.[22] However, many of these potential alternative habitats were also reclaimed for agricultural use.

Land and River Drainage

Throughout the Community wetland areas provide important wildlife habitats. In the British Isles, our mild climate enables our wetlands to support large populations of waterfowl when those at similar latitudes on mainland Europe are frozen.[23] In Ireland, the Shannon Callows alone provide habitats for seven species of bird whose conservation is of international importance: the Greenland white-fronted goose, the whooper swan, the beswick swan, the wigeon, the shoveler, the golden plover and the black-tailed godwit.[24] Additionally, wetlands such as these also act as natural filters, helping to protect water quality.[25]

Wetlands, however, have been the subject of extensive arterial and field drainage. Although this makes the fields more productive, it also reduces their susceptibility to winter flooding and limits feeding opportunities for waterfowl.[26] Drainage also

---

20   Nature Conservancy Council (1984), *Nature Conservation in Great Britain*, (Shrewsbury: Nature Conservancy Council).

21   For example, in a memorandum to the House of Commons Select Committee on the Environment in 1990, the RSPB noted that 'Northern Ireland's hedgerows are still largely intact.' See House of Commons Select Committee on the Environment (1991), *1st Report: Environmental Issues in Northern Ireland*, Session 1990–1991 (London: HMSO) 171.

22   Green, B. Agriculture and the Environment, [1986] 3 *Land Use Policy* 193 at 196.

23   Cadbury, C.J., The Effects of Flood Alleviation and Land Drainage on Birds of Wet Grasslands, in Jenkins D. (ed.) (1984), *Agriculture and the Environment*, (Cambridge: Institute of Terrestrial Ecology).

24   Nairn, R.G.W., Floodplain Agriculture in Ireland and its Significance for Bird Conservation, in the Joint Nature Conservation Committee (1991), *Birds and Pastoral Agriculture in Europe* (Peterborough: Joint Nature Conservation Committee) 93.

25   See, for example, Fennessy, S. (1989), *Riparian Buffer Strips: Their Effectiveness for the Control of Agricultural Pollution*, (London: University College) chapter 2.

26   See, for example, Tucker, G. Priorities for Bird Conservation in Europe: The Importance of the Farmed Landscape, in Pain D., and Pienkowski, M., (eds) (1997),

renders soil-bound insects less accessible to birds[27] and makes it more difficult for wetland plants to survive.[28] Collectively, these changes had a drastically detrimental effect on wetland birds. For example, the RSPB has estimated that drainage on the Somerset Levels in England caused a 69 per cent reduction in bird populations of snipe there between 1977 and 1987.[29] On a European scale, wetland drainage has been identified as being a serious threat affecting some 28 per cent of all birds with unfavourable conservation status.[30] Drainage work has also impacted upon wild plants. The Nature Conservancy Council has previously pointed out that wetland based wild plants make up 69 of an identified 149 species of wild plant in Great Britain whose populations had declined by 20 per cent of more between 1930 and 1984.[31] Elsewhere, arterial drainage schemes also had adverse effects upon fish numbers, by damaging habitats suitable for spawning.[32] It is, perhaps, easy to see why arterial and land drainage have been labelled the most destructive aspects of intensive agriculture affecting nature conservation.[33] In Northern Ireland and the Republic of Ireland, particular criticism has been made of a drainage scheme on the River Blackwater that was jointly instituted by the British and Irish governments. This drainage scheme was conducted under Directive 79/197, which provided Community financial aid to help solve drainage problems being experienced by farmers in border areas.[34] The scheme was principally justified on social grounds – the prevention of rural depopulation, assisting farmers that were farming agriculturally-disadvantaged land and creating employment in areas of high unemployment.[35] One commentator has noted that, in practice, there was

---

Farming and Birds in Europe (London: Academic Press) 105.

27   *Ibid.*

28   Hill C. and Langford, T. (1992), *Dying of Thirst: A Response to the Problem of our Vanishing Wetland* (Lincoln: The Wildlife Trusts Partnership) 8.

29   Robins, M., Davies, S.G.F., Biusson, R.S.K. (1991), *An Internationally Important Wetland in Crisis: The Somerset Levels and Moors – A Case History of Wetland Destruction* (Sandy: Royal Society for the Protection of Birds) 7.

30   Tucker G. and Heath, M. (1994), *Birds in Europe: Their Conservation Status* (Cambridge: Birdlife International) 47.

31   Nature Conservancy Council (1984), *Nature Conservation in Great Britain* (Shrewsbury: Nature Conservancy Council) 64. Of the other plant species that had declined by more than 20 per cent in this time period, 32 belonged to permanent grassland habitats, 18 to woodland habitats and 14 to sandy or heath land habitats.

32   See the Memorandum of the Fisheries Conservancy Board for Northern Ireland to the House of Commons Environment Committee (1991), First Report: *Environmental Issues in Northern Ireland*, Session 1990–1991, (London: HMSO) 143.

33   See, Agra Europe (1991), Special Report No.60: *Agriculture and the Environment – How will the EC Resolve the Conflict* (London: Agra-Europe Ltd) 20 or Newbould, P.J. Nature Conservation and Agriculture, in Jeffrey, D.W. (ed.) (1984), *Nature Conservation in Ireland: Progress and Problems* (Dublin: Royal Irish Academy) 103.

34   Council Directive 79/197, [1979] OJ L43/23.

35   See Newbould, P.J., footnote 33 above, 103.

little economic justification for the work.[36] The environmental impact of the work can be seen from the fact that a 52 per cent fall was, subsequently, recorded in the number of breeding waders present in adjoining farmlands between 1982 and 1992.[37]

Drainage has also had a detrimental impact upon peat lands. This is of particular significance in Ireland, which is traditionally associated with these habitats.[38] Indeed, on a European level, only Finland has a larger percentage surface area of peat.[39] Peat lands in Ireland are of both European and international significance.[40] As the House of Commons Environment Committee has remarked, 'tropical countries have contrasted our own failure to conserve peat bogs with demands upon them to conserve the rain forest.'[41] Just like the rain forests, peat lands play an important climate change role. They act as natural carbon sinks, storing carbon. However, agricultural change, whether by drainage and ploughing or the effects of overgrazing in exposing the peat, releases this carbon into the atmosphere in the form of the greenhouse gas carbon dioxide. In practice, peat lands have been subjected to widespread agricultural drainage. One study, in 1988, estimated that some 15 per cent of upland blanket peat bog in Northern Ireland has been drained

---

36   Milton, K. (1990), *Our Countryside, Our Concern: The Policy and Practice of Conservation in Northern Ireland* (Belfast: Northern Ireland Environmental Link) 43. Milton notes that in Great Britain arterial drainage schemes would only be authorised at that time if they were shown to provide an economic return of 5 per cent or more. (Based upon a cost benefit analysis which weighed the costs of the drainage and farm improvement work against the likely increases in agricultural production.) In contrast, it had been estimated that the River Blackwater drainage scheme would only produce a 3 per cent economic benefit.

37   See Christie, S. (1996), *Environmental Strategy for Northern Ireland* (Belfast: Northern Ireland Environmental Link) 130, or Williams, G., Northern Ireland Drainage Policy, [1988] 9 *Ecos* 43, or Williams, G., Newson, M., and Browne, D. Land Drainage and Birds in Northern Ireland, [1988] 2 *RSPB Conservation Review* 76.

38   Hammond, R.F. (1979), *The Peatlands of Ireland* (Dublin: Soil Survey Bulletin) 35, An Foras Taluntais, estimates that peatlands originally accounted for 17 per cent of the land surface of the Republic of Ireland, whilst Hamilton, A., Peatland, in Cruikshank J.G., and Wilcock, D.N. (eds) (1982), *Northern Ireland: Environment and Natural Resources* (Belfast: Queen's University and New University of Ulster) 185 estimate that they originally accounted for 12 per cent of the land surface of Northern Ireland.

39   Doyle, G.J., Progress and Problems in the Conservation of Irish Peatlands 1983 to 1991, in Feehan, J. (ed.) (1991), *Environment and Development in Ireland* (Dublin: The Environmental Institute, University College Dublin) 511.

40   See Boyle, C.J., Bog Conservation in Ireland, in Schouten, M.G.C. and Nooren, M.J. (eds) (1990), *Peatlands, Economy and Conservation* (The Hague: SPB Academic Publishing) 47.

41   House of Commons Environment Committee (1991), First Report, *Environmental Issues in Northern Ireland*, Session 1990/91 (London: HMSO) xix.

for agricultural use.[42] The land is then ploughed and reseeded, removing most of the flora and fauna that had previously been supported there. Additionally, drainage on agricultural land adjoining peat land has also been a source of damage, by lowering the water table and drying out the peat land.[43] Aside from environmental damage caused by drainage, peatlands have, traditionally, been used for pasture. As such, they have also been subject to environmental damage through overgrazing. One estimate is that overgrazing affected some 40 per cent of blanket bog land in the Republic of Ireland, with the worst affected areas being in the western counties of Galway and Mayo.[44]

Increasing Livestock Numbers

Livestock farming within the Community has experienced a number of changes. On the one hand, the number of livestock farms has fallen, but on the other hand the number of livestock per holding has increased quite dramatically, since the 1970s. The European Commission has reported that the number of farms specialising in livestock farming decreased by one third over the period 1975 to 1995, but that, during this period, the total agricultural area used for this type of farming had decreased by only 6 per cent.[45] In other words, particularly in central and northern areas of the European Community, small extensive farms were being replaced by larger, more intensive farms.

Against this background, the average number of cattle per holding more than doubled between 1975 and 1995.[46] Northern Member States continue to have much higher average livestock numbers per holding than southern Member States.[47] The figures also mask a fall in the number of dairy cattle across the

---

42   Tomlinson, R.W. and Cruickshank, M.M. Monitoring Changes in Hill Peatland in Northern Ireland, in Montgomery, W.I., McAdam, J.H. and Smith, B.J. (eds) (1988), *The High Country: The Land Use and Land Use Change in Northern Irish Uplands* (Belfast: Institute of Biology and Geological Society of Ireland) 59.

43   Baldock, D. (1990), *Agriculture and Habitat Loss in Europe*, (Gland, Switzerland: WWF) 20.

44   Foss, P. National Overview of the Peatland Resource in Ireland, and Bleasadale, A. Overgrazing in the West of Ireland – Assessing Solutions., both in O'Leary, G. and Gormley, F. (eds) (1998), *Towards a Conservation Strategy for the Bogs of Ireland* (Dublin: Irish Peatland Conservation Council).

45   European Commission (1999), *Agriculture, Environment and Rural Development: Facts and Figures – A Challenge for Agriculture* (Luxembourg: Office for Official Publications of the European Communities) 81.

46   *Ibid.* 83.

47   *Ibid.* The European Commission notes, for example, that the average number of cattle per holding in the United Kingdom in 1995 was 87, as against an average of 10 cattle per holding in Portugal.

European Community, following the introduction of milk quotas.[48] Certainly, the introduction of the milk quota provides a practical example of the impact that changes to Common Agricultural Policy production policy can have upon decisions taken by individual farmers. The milk quota was introduced in 1984 to discourage farmers from increasing their production levels.[49] Punitive levies were imposed when milk production exceeded the quota allocated to each farmer. As a result of improvements in technology, the rearing of more productive cattle and better awareness of the impact of animal feedstuffs on production, milk productivity per cow has steadily increased.[50] In the period 1975 to 1995, average milk production per dairy cow increased by 45 per cent.[51] The increased productivity also continued beyond the introduction of milk quotas in 1984, with an increase of 20 per cent being recorded in the Community between 1985 and 1995.[52] As a result of the introduction of milk quotas, and in light of the increased productivity being achieved, dairy cattle numbers fell across the European Community.[53] Equally, as noted above, land formerly used as pasture for dairy cattle has been transferred to arable cropping.

Increases in the numbers of sheep farmed within the European Community provide a further example of the influence of Common Agricultural Policy measures. In 1980 the European Community introduced a common organisation in sheepmeat, which provided for a sheep annual premium payment to be made to farmers when average market prices were below a basic price set by the Community.[54] The result was that sheep numbers soared by almost 50 per cent within the EU in the period to 1992.[55] The average number of sheep per hectare also increased sharply over this period.[56] In 1993 the European Community introduced livestock quotas, to limit farmers' eligibility for sheep annual premium. This, effectively, acted to stabilise sheep numbers.[57]

At farm level, increased livestock numbers pose a potential environmental problem. For example, the probability of ground nesting birds being able to safely hatch their eggs is reduced. Studies have shown that approximately 40 per cent of lapwing nests are trampled at stocking levels of one cow per acre, whilst for

---

48   European Commission, footnote 45 above. Though, exceptionally, the European Commission notes that both Greece and Italy have continued to record increased cattle densities upon farms.

49   By virtue of Council Regulation 857/84, [1984] OJ L90/13.

50   European Commission, footnote 45 above, 87.

51   *Ibid.* Based upon the 9 Member States who were European Community members throughout this period.

52   *Ibid.* Based upon the 10 Member States who were European Community members throughout this period.

53   *Ibid.* p.86.

54   By virtue of Council Regulation 1837/80, [1980] OJ L183/1.

55   European Commission, footnote 45 above, 93.

56   *Ibid.*

57   *Ibid.*

snipe and redshank the figures are 60 and 72 per cent respectively.[58] However, at double that stocking level these birds lose respectively 60, 85 and 93 per cent of their nests.[59]

Sheep farming is a distinctive feature of upland farming in Great Britain, where heather moorland provides a habitat for a wide range of insects, mammals and birds.[60] This habitat, however, has been threatened by increased sheep numbers. The appropriate grazing density of heather alters from site to site. However, it has been suggested that heather will decline in both condition and extent when sheep numbers exceed an average of 1.5 to 2.0 sheep per hectare.[61] Research in England suggests that average stocking levels in English uplands in the early 1990s often exceeded such optimum rates.[62] The ultimate effect of overgrazing upon upland farms is to replace heather moorland with either bare soil or, in due course, natural grasses that are often of little ecological or agricultural value.[63] This situation can be compounded by a number of additional factors. Traditionally, farmers would shepherd their sheep, so that the uplands were evenly grazed. However, reductions in the farm labour force mean that this often no longer occurs and sheep graze at will, causing damage by overgrazing in particular areas.[64] Additionally, the practice of providing supplementary feed sites for upland sheep encourages sheep

---

58   O'Connor, R.J. and Shrubb, M. (1986), *Farming and Birds* (Cambridge: Cambridge University Press) 161.

59   *Ibid.*

60   See Thompson, D.B.A., McDonald, A.J., Marsden, J.H., Galbraith C.A. Upland Heather Moorland in Great Britain: A Review of International Importance, Vegetation Change and Some Objectives for Nature Conservation, [1995] 71 *Biological Conservation* 163.

61   Bardgett, R.D., Marsden, J.H., Howard, D.C. The Extent and Condition of Heather on Moorland in the Uplands of England and Wales, [1995] 71 *Biological Conservation* 155. For other estimates of optimum stocking rates to avoid heather loss and damage see Thompson, D.B.A. et al., footnote 60 above at 173, Andrews, J. and Rebane, M. (1994), *Farming and Wildlife – A Practical Management Handbook* (Sandy: RSPB) and Evans, R. (1996), *Soil Erosion and its Impact in England and Wales* (London: Friends of the Earth) 24.

62   Council for the Protection of Rural England (1994), *Down to Earth: Environmental Problems Associated with Soil Degradation in the English Landscape* (London: Council for the Protection of Rural England) 17. For earlier research see also Sinclair, G., *The Uplands Landscape Survey* (Dyfed: Environment Information Services).

63   See White, B. Natural Resource Management: The Case of Heather Moorland, in Allanson, P. and Whitby, M. (eds) (1996), *The Rural Economy and the British Countryside* (London: Earthscan Publications Ltd) 69.

64   See, The Wildlife Trusts (1996), *Crisis in the Hills: Overgrazing in the Uplands* (Lincoln: The Wildlife Trusts) 6, Lovegrove, R., Shrubb, R., Williams, I. (1995), *Silent Fields – The Current Status of Farmland Birds in Wales* (Sandy: RSPB) 9 or Felton, M., Marsden, J. (1990), *Scientific and Policy Initiatives: Heather Regeneration in England and Wales* (Peterborough: Nature Conservancy Council) 16.

to spend long periods in the vicinity of those feed sites.[65] This again encourages damage by overgrazing in these areas. In 1990 it was estimated that half of England's heather moorland was in a poor or suppressed condition.[66] Similar studies have been conducted in Ireland. In one study it was shown that, as result of overgrazing, Hen Mountain in Northern Ireland's Mourne Mountains had lost 64 per cent of its heather moorland in just ten years.[67] Equally, in the Republic of Ireland overgrazing has been linked to the destruction of large areas of heather moorland in Connemara, western Ireland.[68] Given that it can take up to fifty years to successfully re-establish damaged heather,[69] these situations give cause for concern.

Increased Use of Agricultural Technology

Increased use of technology in agricultural practice can be gauged in several ways. In this section it will be considered from the perspective of increased use of artificial fertilisers and chemical pesticides. In Europe as a whole, not just the European Community, average consumption of nitrogen based artificial fertilisers increased by 75 per cent between 1970 and 1989.[70] Within the European Community, it was reported in 1991 that artificial fertiliser usage in north-western Europe was the highest in the world.[71] Nitrogen based fertiliser is, today, commonly applied to

---

65   Lovegrove. R., Shrubb, R. and Williams, I. footnote 64 above, 9.

66   Nature Conservancy Council (1990), *Scientific and Policy Initiative: Heather Regeneration in England and Wales* (Peterborough: Nature Conservancy Council). Other studies have sought to estimate the degree of heather loss in particular areas. Anderson, P. and Yalde, D.W. Increased Sheep Numbers and Loss of Heather Moorland in the Peak District, England, [1981] 20 *Biological Conservation* 195, estimated that 36 per cent of the Peak District's upland heather had been lost in the period between 1914 and 1979. Felton, M. and Marsden, J.H. (1990), *Heather Regeneration in England and Wales: A Feasibility Study for the Department of the Environment* (Peterborough: Nature Conservancy Council), estimated that Cumbria had lost 36 per cent of its heather moorland in the period 1940–1980.

67   Christie. S. (1996), *Environmental Strategy for Northern Ireland* (Belfast: Northern Ireland Environmental Link) 129.

68   See Bleasdale, A., Sheey-Sheffington, M. The Influence of Agricultural Practices on Plant Communities in Connemara, in Feehan, J. (ed.) (1991), *Environment and Development in Ireland* (Dublin: The Environmental Institute, University College) 331.

69   White. B. Natural Resource Management: The Case of Heather Moorland, in Allison. P. and Whitby, M. (eds) (1996), *The Rural Economy and the British Countryside* (London: Earthscan Publications Ltd) 78.

70   Stanners, D., Bourdeau, P. (eds) (1995), *Europe's Environment: The Dobřiš Assessment* (Copenhagen: European Environment Agency) 455.

71   Agra-Europe Ltd (1991), *Agriculture and the Environment: How will the EC Resolve the Conflict?* (London: Agra-Europe Ltd) 12.

crops and grassland.[72] The water pollution problems associated with this fertiliser use will be examined below. Additionally, however, its impact upon biodiversity should also be considered. Artificial fertilisers are used because of their potential to boost productivity amongst target crops or grasses. However, they also have a negative impact upon wild plants with lower nutrient tolerances. A two-year study revealed that the application of 200 kg of nitrogen based fertiliser per hectare upon hay meadows reduced the species diversity recorded there by 15 per cent.[73] Even lower dosages, of 50 kg of per hectare, were shown to have adverse effects by reducing botanical diversity.[74] A strong link has also been established between high nitrogen usage and declining populations of large insects in grasslands.[75] This decline has been linked to fertiliser usage of over 50 kg per hectare.[76] In turn, the insect decline can also be linked to declining bird populations, particularly in wader species such as lapwing and curlew for whom these insects are an important food source.[77]

The use of chemical pesticides also poses a major potential threat to biodiversity. Over 800 chemical pesticides are registered for use within the European Community.[78] In recent years levels of pesticide use have actually been falling within the European Community, but to some extent this has been

---

72 O'Connor, R.J., Shrubb, M. (1986), *Farming and Birds* (Cambridge: Cambridge University Press) 214 estimate that in England and Wales in 1982 nitrogen based fertilisers were used on 96 per cent of arable land, 93 per cent of improved grasslands and 77 per cent of unimproved grasslands. See also Fuller, R.M. The Changing Extent and Conservation Interest of Lowland Grasslands in England and Wales: A Review of Grassland Surveys 1930–1984, [1987] 40 *Biological Conservation* 281 at 291.

73 Tallowin, J., Mountford, O., Kirkham, F. Fertilisers on Hay Meadows: a Compromise? [1994] 2 *Enact* 16. They also noted that after seven years of fertilisation at this rate, botanical diversity had decreased by 40 per cent. See also Wells, T.C.E. Responsible Management for Botanical Diversity, in British Grassland Society (1989), *Environmentally Responsible Grassland Management* (Hurley: The British Grassland Society) 44 and Baldock, D. (1994), *The Nature of Farming: Low Intensity Systems in Nine European Countries* (London: Institute for European Environmental Policy) 13.

74 Tallowin, J. et al., footnote 73 above. Over a seven-year period, at this dosage rate, a 20 per cent reduction in botanical diversity was recorded.

75 Beintera, A.J., Thissen, J.B.M., Tensen, D., Visser, G.H. Feeding Ecology of Charadriform Chicks in Agricultural Grassland. [1991] 79 *Ardea* 31 and also Lovegrove, R., Shrubb, M., Williams, I. (1991), *Silent Fields: The Current Status of Farmland Birds in Wales* (Sandy: RSPB) 10.

76 Beintera A.J. et al., footnote 75 above.

77 Beintera, A.J. Insect Fauna and Grassland Birds, in Joint Nature Conservation Committee (1991), *Birds and Pastoral Agriculture in Europe* (Peterborough: Joint Nature Conservation Committee) 97.

78 European Environment Agency (1998), *Europe's Environment: The Second Assessment* (Copenhagen: The European Environment Agency) 187.

compensated for by the use of more concentrated products.[79] The potential scale of the overuse of pesticide products is revealed by studies that showed that farmers in the Netherlands and Germany had been able to reduce their pesticide usage by 36 per cent and 60–90 per cent, respectively, without experiencing any income loss.[80]

The effects of pesticides upon biodiversity have been known in Great Britain since the early 1960s when the deaths of seed eating birds, such as wood pigeons, pheasants and rooks, together with foxes and cats were traced to the practice of dressing seeds with highly toxic pesticides.[81] Subsequently, in the United States, Rachel Carson's polemic book *Silent Spring* drew attention to the harmful effects of pesticides such as DDT.[82] In particular, this book revealed how toxic these chemicals were, in the sense that they remained active within the food chain and caused death or sterility to large numbers of wildlife. Today, pesticides such as DDT are banned and have been replaced by pesticides that decompose more rapidly and, therefore, will not multiply within the food chain.[83] Nevertheless, the use of any pesticide still carries a risk of environmental side-effects. [84] For example, the application of herbicide, as would be expected, is a successful means to control weeds amongst crops.[85] However, it is also likely to substantially reduce the population of fauna that is dependent upon those plants for food or shelter. Insects are particularly vulnerable.[86] A reduction in insect numbers, either as the indirect result of the use of herbicide or as the direct result of the targeted use of insecticides, also impacts upon bird life and other animals that feed upon these insects. One example is that high mortality amongst grey partridge chicks has been

---

79   European Environment Agency (1995), *Environment in the European Union 1995* (Copenhagen: European Environment Agency) 81.

80   The Royal Commission on Environmental Pollution, 19th Report, *Sustainable Use of Soil*, Cm. 3165 para. 5.80.

81   See Sheail, I. (1985), *Pesticides and Nature Conservation* (Oxford: Clarendon Press Ltd).

82   Carson, R. (1962), *Silent Spring* (United States: Houghton Mifflin).

83   See, for example, the Royal Commission on Environmental Pollution, footnote 80 above at paragraph 5.80.

84   For example, it has been reported that pesticides were responsible for a third of all animal poisoning deaths in the United Kingdom in 1996. See, European Commission (1999), *Agriculture, Environment, Rural Development: Facts and Figures – A Challenge for Agriculture* (Luxembourg: Office for Official Publications of the European Communities) 190.

85   See, for example, Marshall, J. Weeds, in Greg-Smith, P., Frampton, G. and Hardy, T. (eds) (1984), *Pesticides, Cereal Farming and the Environment: The Boxworth Project* (London: HMSO) 25.

86   See, for example, Vickerman, G.P. The Effects of Different Pesticides on the Invertebrate Fauna of Winter Wheat, in Greg-Smith, P. et al., footnote 85 above, 82.

correlated with increased pesticide use, which has reduced their food source.[87] Similarly, insecticides can also eradicate insects such as bees whose existence is vital for the pollination of other plants or crops.[88]

The environmental effects of pesticides are also influenced by the manner in which they are used. In most cases, pesticides will be sprayed from tractor mounted units. In this situation, it is not uncommon for some of the spray to be caught by the wind and blown onto unintended areas such as hedges or field margins. This drift will have a detrimental impact upon flora and fauna in these areas. Estimates vary as to the distance over which this affect will be felt.[89]

One further example of the environmental impact of developments in agricultural technology has been the move from hay making to silage production upon grassland farms today. Silage tends to be cut earlier than hay. In particular, this impacts upon ground nesting birds. One species that has been particularly badly affected has been the corncrake, the only bird species found upon the island of Ireland, that is threatened with global extinction.[90] The impact of the earlier cutting date is that when silage cutting takes place nests containing eggs may be destroyed.[91] It has also been a cause of deaths amongst adult corncrakes.[92] Studies in Great Britain have shown a 75 per cent reduction in corncrake numbers between 1970 and 1990.[93]

---

87   See Potts, D. Cereal Farming, Pesticides and Grey Partridges, in Pain, D.J. and Pienkowski, M. (eds) (1997), *Farming and Birds in Europe: The Common Agricultural Policy and its Implications for Bird Conservation* (London: Academic Press) 152. See also Sotherton, N.W. Conservation Headlands: A Practical Combination of Intensive Cereal Farming and Conservation, in Firbank, L.G., Carter, N., Darbyshire, J.F. and Potts, G.R. (eds) (1991), *The Ecology of Temperate Cereal Fields: 32nd Symposium of the British Ecology Society* (Oxford: Blackwell Scientific Publications) 384.

88   See Pimenta, D. et al., Environmental and Economic Costs of Pesticide Use, [1992] 42 *Bioscience* 750 at 754.

89   For example, Mars, R.H., Frost, A.J., Plant, R.A. and Lunnis, P. Effects of Herbicides on Vegetation, in English Nature (1993), *The Environmental Effects of Pesticide Drift* (Peterborough: English Nature) 35, estimate that 'for established perennial species severe damage from herbicide drift was confined to within 10 metres downwind...' In contrast, in the same volume, Davies, B.N.K. et al., *Effects of Insecticides on Terrestrial Invertebrates*, at 61, report that more toxic insecticides may be effective for 10 to 24 metres from the spraying unit.

90   Collar N.J., Crosby M.J. Statterfield A.J. (1994), *Birds to Watch 2: The World List of Threatened Birds* (Cambridge: Birdlife International), 77. Batten, L.A., Bibby, C.J., Clement, P., Elliot, G.D., Potter, R.F. (1990), *Red Data Birds in Britain: Action for Rare and Important Species* (London: Nature Conservancy Council, and Sandy, RSPB) 135.

91   O'Connor R.J., Shrubb, M. (1986), *Farming and Birds* (Cambridge: Cambridge University Press) 93.

92   See, for example, Derwin, J. (1997), *Corncrake Fieldwork in North Donegal* (Dublin: Birdwatch Ireland) 14.

93   Fuller, R.J. et al. (1995), Population Declines and Range Contractions Among Lowland Farmland Birds in Britain, [1995] 9 *Conservation Biology* 1425 at 1428.

Agriculture's Impact on Biodiversity, Water Pollution and Soil Erosion

It is clear that agricultural production policies, both under the Common Agricultural Policy and the national policies that preceded it, have had an important influence upon agricultural practices within the British Isles and the European Community as a whole. Equally, these developments in agricultural practice have been purchased at a cost to the wider environment. This section will now explore that environmental cost. It will examine the role that agriculture has played in relation to three particular environmental problems – biodiversity loss, water pollution and soil erosion.

Agriculture and Biodiversity Loss

Each of the changes in agricultural practice, that have been highlighted above, has had negative effects on biodiversity. Overall, in Great Britain it has been reported that in the period 1945–1984 10 species of flowering plant, 3 or 4 species of dragonfly and 1 butterfly have become extinct.[94] These extinctions have been linked to the land use changes associated with agricultural intensification.[95] Additionally, in the same time period, it has been reported that some 149 plants, 11 species of dragon fly, 13 species of butterfly, 36 species of birds, 4 species of reptile or amphibian and several species of mammal, in particular otters and bats, have sustained serious population declines.[96] Today, the danger of species extinction is no less real. For example, in 1998, WWF predicted that more wildlife would continue to become extinct over the next 20 years as a direct consequence of agricultural practices.[97]

A different threat to biodiversity stems from the fall that has occurred in the number of farmland animal breeds. In the period since 1892, 26 breeds of farmland animal have become extinct in the United Kingdom.[98] Local breeds throughout Europe have been improved by selected cross-breeding or displaced by other

---

94 Nature Conservancy Council (1984), *Nature Conservation in Great Britain* (Shrewsbury: Nature Conservancy Council) 61–64.

95 *Ibid.* See also Green, B. Agriculture and the Environment: A Review of Major Issues in the United Kingdom, [1986] 3 *Land Use Policy* 193.

96 Nature Conservancy Council, footnote 94 above. In respect of bird populations in serious decline see also Fuller, R.J., et al. Population Declines and Range Contractions Amongst Lowland Farmland Birds in Britain, [1995] 9 *Conservation Biology* 1425 and O'Connor, R.J., Shrubb, M. *Farming and Birds* (Cambridge: Cambridge University Press) 149–186.

97 WWF, *Doomsday for Nature*, WWF Press Release 14th December 1998. The particular species that they identified were the high brown fritillary butterfly, the marsh fritillary butterfly, the song thrush, the skylark and the grey partridge.

98 World Conservation Monitoring Centre (1992), *Global Diversity: the Status of the Earth's Living Resources* (London: Chapman and Hall) 401.

more productive breeds.[99] For example, 95 per cent of the Community's dairy herd today is composed of just two breeds – Holstein and Friesian.[100] Although these breeds are more productive they also require more inputs in terms of feeding and management and are often not as suited to prevailing physical conditions.[101] The end result is that some traditional breeds have been lost whilst others have dwindled in numbers. A similar situation exists in relation to crops. One estimate states that farmers now rely on just nine species of plant for 75 per cent of crops.[102] This creates a potential risk of crop failure should these species become affected by disease.

Water Pollution

Agriculture has also been a major cause of water pollution. In particular, mismanagement of slurry and silage effluent has led to pollution incidents, as has the drainage of soiled waters from farmyards into groundwaters or neighbouring water courses. The National Rivers Authority has previously suggested that as many as 40 per cent of farms within each river catchment may either already be contributing to such water pollution or be at high risk of doing so.[103] In these situations, agricultural pollution provides a source of food for micro-organisms within the water, which in turn expend the water's oxygen supply whilst breaking it down. The impact of this pollution, known as the Biological Oxygen Demand, will vary depending upon the particular source of the waste. Typical examples are shown in Table 2.1.

In serious cases, the reduction in water oxygen levels results in fish deaths. In less serious cases, it is likely to reduce the chemical and biological quality of individual watercourses.

Increased use of nitrogen based artificial fertilisers also had consequences for water pollution. Overall, it has been estimated that, on average, a surplus of between 50–100 kg of nitrogen per hectare is applied to agricultural soils within the European Community each year.[104] In practice, plants only take up nutrients in accordance with their own needs. Excess nutrients, not taken up by the plants, remain in the soil and are washed into adjacent water bodies as run-off or leech into groundwaters. Indeed, a substantial proportion of fertiliser applications

---

99   See Baldock, D. (1994), *The Nature of Farming: Low Intensity Farming in Nine European Countries*, (London: Institute for European Environmental Policy) 46.

100   Agra-Europe Ltd (1991), *Agriculture and the Environment – How will the EC Resolve the Conflict?* (London: Agra Europe Ltd) 19.

101   *Ibid.*

102   Emerson C., Jenkins R. (1995), *Thought for Food* (London: SAFE Alliance) 2.

103   National Rivers Authority (1992), *The Influence of Agriculture on the Quality of Natural Waters in England and Wales* (Almondsbury: National Rivers Authority) 20.

104   European Environment Agency (2003), *Europe's Water: An Indicator Based Assessment*, (Copenhagen: European Environment Agency) 44.

Table 2.1   Water Pollution: Rates of Biological Oxygen Demand

Waste	Biological Oxygen Demand (mg/l)
Treated Sewage	20–60
Raw Domestic Sewage	300–400
Dilute Dairy Parlour/Yard Washings	1,000–5,000
Liquid waste draining from slurry stores	1,000–12,000
Liquid Sewage Sludge	10,000–20,000
Cattle Slurry	10,000–20,000
Pig Slurry	20,000–30,000
Silage Effluent	30,000–80,000
Milk	140,000

Source: © Crown copyright 1998. Originally published in the *Code of Good Agricultural Practice for the Protection of Water* (The Water Code), Ministry of Agriculture, Fisheries and Food, 1998, p. 2. Reproduced with kind permission of DEFRA.

wash directly into adjacent water courses as run-off.[105] Overall, the European Environment Agency, in a comparison of the periods 1977–1992 and 1988–1990, reported that increased nitrate levels were recorded in two thirds of European rivers.[106]

Nitrogen based fertilisers are a source of nitrate, a compound of nitrogen. Within the European Community concern about nitrate levels within European waters raised public health concerns. Nitrate pollution has been linked to the condition methaemoglobinaemia, also known as blue baby syndrome, which is potentially fatal.[107] It has also been suggested that it may have an influence upon human cancers, though no direct link has been scientifically established.[108] Such

---

105   Maitland, P.S., The Effects of Eutrophication on Aquatic Wildlife, in Jenkins, D. (ed.) (1984), *Agriculture and the Environment* (Cambridge: Institute of Terrestrial Ecology) 102. Maitland refers to several studies that have shown that between 10 and 25 per cent of nitrogen applied as fertiliser runs off into adjacent watercourses. See also Dietz, F.J., Heijnes, H. Nutrient Emissions From Agriculture, in Dietz, F.J., Vollebergh, H.R.J., Varies, J.L. (eds) (1993), *Environment, Incentives and the Common Market*, (Dordrecht, the Netherlands: Kluwer Law International).

106   Stanners, D., and Bordeau, P. (1995), *Europe's Environment: The Dobris Assessment* (Copenhagen: European Environment Agency) 94.

107   See, for example, Dudley, N. (1991), *Nitrates: The Threat to Food and Water* (London: Merlin Press Ltd).

108   See, for example, the Royal Commission on Environmental Pollution (1979), 7th Report: *Agriculture and Pollution* (London: HMSO) 212.

health concerns induced the European Community to introduce a requirement that nitrate concentrations within drinking waters should not exceed 50 mg/l. [109] The Community also encouraged Member States to achieve a voluntary guideline concentration of 25 mg/l.[110] In 2003, however, the European Environment Agency estimated that the maximum concentration level was being exceeded in around one third of the groundwater bodies for which information was available.[111] Thus, given that groundwaters are an important source of drinking waters across Europe, nitrate pollution remained an important problem.

Increased nitrate levels also contribute towards eutrophication in European waters. Eutrophication is, essentially, the enrichment of individual waterways by nutrients, such as nitrate and phosphate, that encourage algal growth. This leads to the development of algal blooms, particularly on the surface of lakes or slow moving rivers. By reducing light penetration and oxygen levels within the water, these algal blooms reduce the diversity of plant and fish life that can survive. They also have an important human impact, in that some algae contain toxins that can cause illness through direct contact in the water or their accumulation in sea-food.[112] Their appearance is often unsightly and can also be associated with an unpleasant smell. This can have a negative impact upon regional economies by discouraging recreational use of these waters. The European Environment Agency has identified eutrophication as being a widespread environmental problem in European Community water ways.[113] Agriculture and discharges made by sewage treatment plants have been identified as being the principal sources of the water pollution providing the additional phosphate and nitrate that triggers eutrophication.[114] In the case of agriculture, it has been the use of animal manures and artificial fertilisers that have been the main sources of these nutrients.[115]

In seeking to limit the levels of nitrate found in particular water bodies, Member States have the added problem that it can take a substantial period of time before nitrates leech into surface waters. It can take 20 or more years before this

---

109   Council Directive 75/440 [1975] OJ L194/26 on drinking water quality.

110   *Ibid.*

111   European Environment Agency (2003), *Europe's Environment: The Third Assessment* (Copenhagen: European Environment Agency) 172.

112   See, for example, Klein, G., Perera, P. (2000), *Eutrophication and Health,* (Copenhagen: European Environment Agency).

113   See, generally, Stanners D., Bordeau, P. footnote 106 above, 89–94 and 571–572, The European Environment Agency (1998), *Europe's Environment: The Second Assessment* (Copenhagen: European Environment Agency) 193–202 and 210–214 and The European Environment Agency (2003), *Europe's Environment, the Third Assessment* (Copenhagen: European Environment Agency) 174–184.

114   See European Environment Agency (2000), *Nutrients in European Ecosystems* (Copenhagen: European Environment Agency) 21.

115   *Ibid.*

process is complete.[116] For example, increased nitrate levels in waterways in areas of England and Wales in the 1970s have been linked to intensification in grassland use in the 1940s.[117]

Pesticide use in agriculture also raises water pollution concerns. Public concern over the use of pesticides in agriculture mainly centres upon the risk of them entering the food chain through contaminated agricultural produce or public water supplies. The European Community has set quality standards for pesticide levels in drinking waters. Under Directive 80/778 the maximum permissible concentration has been set at 1 part per billion for individual pesticides.[118] One estimate, however, suggested that this level was exceeded in groundwaters beneath 75 per cent of the arable land in the European Community.[119] Additionally, research in Great Britain has shown that water pesticide levels tend to be much higher in ditches draining from farmyards, rather than from fields.[120] This would suggest that spillages are an important pollution source. This research would also raise concerns about the quality of waters obtained from bore holes situated near farms. The same concern can also be raised in relation to the nitrate content of such bore holes. An examination of rural wells in Belgium found that 29 per cent, of 5000 wells examined, had a nitrate content which exceeded the 50 mg/l limit established by the drinking water directive.[121]

Soil Erosion

Upland overgrazing and the poaching of soils near secondary feeding points has been a cause of soil erosion. These practices lead to the soils becoming exposed to wind and water. The principal types of soil erosion experienced in Great Britain are the formation of sheep scars, by sheep scratching hollows for shelter, the uncovering of rock outcrops through loss of vegetation and soil, gully erosion caused by the channelling of water run-off and sheet erosion caused by the washing away of soils when there is excessive run-off.[122] One survey, in the Peak

---

116   Agra Europe (1991), *Agriculture and the Environment: How Will the EC Resolve the Conflict?* (London: Agra Europe Ltd) 16.

117   National Rivers Authority (1992), *The Influence of Agriculture on the Quality of Natural Waters in England and Wales* (Aldmonsbury: National Rivers Authority) 45. See also Schneider, A. et al. (1999), *Groundwater Quality and Quantity in Europe* (Copenhagen: European Environment Agency) 19.

118   Council Directive 98/83, [1998] OJ L330/32. This has been likened to the equivalent of one drop of concentrated pesticide in an Olympic-size swimming pool, see The Pesticides Safety Directorate (1995), *Keeping Pesticides Out of Water* (London: Ministry of Agriculture Fisheries and Food).

119   Stanners D., Bourdeau, P. footnote 106 above, 72.

120   The Pesticide Safety Directorate, footnote 118 above, 2.

121   European Environment Agency, footnote 104 above, 46.

122   Taylor, A. (1995), *Environmental Problems Associated with Soil in Britain: A Review* (Perth: Scottish Natural Heritage) 36.

District, found that 35 per cent of upland bare soils, susceptible to erosion, were the result of sheep scars and 25 per cent were caused by the removal of heather caused by overgrazing and wind action.[123] As sheep numbers increased, so the levels of soil erosion being experienced in the uplands also increased.[124] Aside from its environmental impact, soil erosion also has economic consequences for farmers, since areas suffering from soil erosion will be less productive

Soil erosion occurs through a combination of wind action and rainfall. Winds will carry off light soil particles, with exposed upland soils being particularly at risk due to higher wind speeds and the fact that these soils are likely to exposed throughout the year.[125] Rain also an important impact in that rain droplets will dislodge soil particles and when the soil becomes saturated with rainfall, these particles will be carried away in sheets of water that form as a result of subsequent rainfall.[126]

However, soil erosion is not just an upland problem. In the lowlands, soil erosion is associated with arable cropping. For example, winter crops and other slower-growing crops take longer to cover and protect the soil.[127] Crop management techniques have also exacerbated soil erosion problems. For example, the removal of hedges, to allow machinery to operate more efficiently, often also removed protection from soil run-off.[128] Also, for safety and convenience most fields tend to be worked in an up-slope-down-slope direction with the result that 'tram-lines' of uncultivated soil between lines of crop and tracks of soil compacted by farm machinery face down slope and promote down-slope soil run-off within rainwater.[129] This in turn causes farmers to lose fertile top-soil and also to lose the benefits of inputs such as fertilisers and pesticides that are also washed away.

Estimates of soil loss vary considerably across the United Kingdom. In areas seriously affected by soil erosion, the erosion can pose a threat to productivity. For example, the Royal Commission on Environmental Pollution has reported that cultivated areas in the Fens of East Anglia are losing soil at a rate of 1cm per annum and predicted that many areas may be 'worked out' by the middle of the century.[130] In Great Britain as a whole, it has been estimated that soil erosion costs

---

123   Evans, R. Overgrazing and Soil Erosion on Hill Pastures With Particular Reference to the Peak District, [1977] 32 *Journal of the British Grassland Society* 65 at 68. The remaining causes of exposed soils were the development of paths/tracks (18 per cent), heather burning (11 per cent) and gravity scars on slopes greater than 35 degrees (8 per cent).

124   Taylor, A., footnote 122 above, 35.

125   See Evans, R. (1996), *Soil Erosion and its Impact in England and Wales* (London: Friends of the Earth) 23.

126   *Ibid.*

127   *Ibid.* See also Royal Commission on Environmental Pollution (1996), 19th Report: *Sustainable Use of Soil* (London: HMSO) 56.

128   Evans, R., footnote 125 above, 30.

129   *Ibid.*

130   Royal Commission on Environmental Pollution, footnote 127 above 57.

the arable industry £2.14 million per annum in lost production and inputs.[131] In addition, the cost of removing deposits of eroded soils from roads, ditches and urban areas in the most severely affected areas of England has been estimated at an additional £2.4 million.

Competing Pressures on Agricultural Policy Reform

In recent years, the Common Agricultural Policy has undergone a number of legislative reforms. In the 1980s, a series of piecemeal reforms were introduced within individual common organisations. More recently, three substantive reform packages have been introduced. These began with the 'MacSharry reforms' in 1992 and also included the 1999 'Agenda 2000' reforms, the 2003 'Mid Term Review of the CAP' and the 2008 'CAP Health Check'.[132] Although agriculture has clearly had a major impact upon the European environment, environmental protection has been only one of several issues that competed to influence these recent policy developments.[133] This policy competition has, inevitably, had an impact upon the manner in which environmental issues have been tackled.

One of the most pressing economic issues affecting the initial Common Agricultural Policy was the fact that increased agricultural production led to the Community moving beyond high levels of self-sufficiency to a position in which many agricultural sectors were producing surpluses. For example, five common organisations (those concerned with the production of milk, beef and veal, cereals, pig meat and fresh vegetables) today account for over 50 per cent of the value of the Community's entire agricultural production.[134] By the mid-1980s each of these common organisations was producing surpluses, as also were those concerned with sugar, wine, eggs and poultry.[135] These production surpluses caused adverse publicity, as attention was drawn to beef and butter 'mountains' and wine 'lakes' that were a by-product of the purchase of surplus produce by national intervention agencies.

Production surpluses also had important consequences for the Community budget. The commitment given by the Community to reimburse national

---

131   Evans, R., footnote 125 above, 43–44.

132   See Chapter 3.

133   See Jack, B. Economy and Environment: Shaping the Development of European Agricultural Law, [2001] 2 *Web Journal of Current Legal Issues*.

134   European Commission (2008), *The Agricultural Situation in the European Union: 2007 Report* (Luxembourg: Office for Official Publications in the European Communities) table 3.1.1. [on-line]. http://ec.europa.eu/agriculture/publi/agrep/index_en.htm [last accessed 14 May 2008].

135   See European Commission (1987), *The Agricultural Situation in the European Community: 1986 Report* (Luxembourg: Office for Official Publications in the European Communities) 344.

intervention agencies, for the costs they incurred in the intervention purchasing of agricultural produce and payment of export refunds, meant that this cost accrued to the Community budget. By 1985, these payments were absorbing 70 per cent of the entire Community budget.[136] Indeed, in the period 1973–1988 they never accounted for less than 63 per cent of all Community budgetary spending.[137] Expanding agricultural production resulted in Community expenditure increasing more quickly than budgetary receipts. In 1985, the Member States agreed to raise their contributions to the Community budget from 1 to 1.4 per cent of national VAT revenues.[138] However, by 1986 even this additional expenditure had been consumed.[139] In response to this situation both the United Kingdom and the Netherlands refused to agree additional funding for the Community budget until the issue of agricultural overproduction had been tackled.[140]

Pressure for reform also came from international competitors. This was evident at the Uruguay Round of trade negotiations within the, then, General Agreement on Tariffs and Trade, held between 1986 and 1994. At these talks other agricultural exporting nations sought radical reductions in the levels of export refunds and domestic agricultural support provided by the Community.[141] In addition, they sought greater access to Community markets, through the removal of variable import levies and the limitation in the use of the principle of Community preference.[142] Ultimately, under the terms of the 1994 GATT Agriculture Agreement, the Community agreed to reduce levels of domestic agricultural support by 20 per cent. In return, direct payments which were independent of production levels were exempted from this requirement. The Community also agreed to cut subsidised exports by 36 per cent in value and 20 per cent in volume. In addition, the Community agreed to amend all import restrictions to fixed customs duties and to reduce these duties by 36 per cent.[143] Cconsequently, these terms also had an important influence on the future direction of the Common Agricultural Policy.

The 1999 and 2003 reforms of the Common Agricultural Policy were similarly motivated by a number of economic considerations. Firstly the Community was

---

136 European Commission (1986), *The Agricultural Situation in the European Community: 1985 Report*, (Luxembourg: Office for Official Publications in the European Communities) 261.

137 Swann D. (1995), *The Economics of the Common Market* (London: Penguin Books Ltd) 260.

138 Council Decision 85/257, [1985] OJ L128/15.

139 Bladen-Howell R., Symons S. The EC Budget, in Artis M., Lee N. (eds) (1991), *The Economics of the European Union* (Oxford: Oxford University Press) 371.

140 Swann, D., footnote 137 above, 260.

141 Ingersent, K.A., Rayner, A.J., Hine, R.C. Agriculture in the Uruguay Round: An Assessment, in Ingersent, K.A., Rayner, A.J., Hine, R.C. (eds) (1994), *Agriculture in the Uruguay Round* (London: Macmillan) 260.

142 Josling, T., Tangerman, S., Warley, T. (1996), *Agriculture in the GATT* (London: Macmillan) 179.

143 *Ibid.* chapter 8: The Uruguay Round Agreement on Agriculture.

about to engage in a further round of international trade negotiations. Under the 1994 GATT Agreement GATT had been replaced by the World Trade Organisation (WTO).[144] The GATT Agriculture Agreement made provision for a further round of negotiations in relation to international agricultural trade to begin in 1999. Article 20 of the agreement provided that the rationale for these negotiations was that 'the long term objective of substantial progressive reductions and protection resulting in fundamental reform is an ongoing process.'

Levels of domestic agricultural support in the European Community were still high in comparison with those of our international competitors. Calculations of domestic support, known as Producer Support Estimates (PSEs) are calculated as a percentage of the value of agricultural production. By way of analogy, in 1998 average PSE's in Australia and New Zealand were 7 and 1 per cent, respectively, whilst in the European Community and the United States they were 45 and 22 per cent, respectively.[145] The prospect of international pressure for further agricultural trade liberalisation was, therefore, contributory to the 1999 reforms of the Common Agricultural Policy.

In recent years, the Community has enlarged to incorporate several new member states, including ten central and eastern European countries.[146] Agriculture, generally, forms a larger part of the economies of many of these new central and eastern European Member States countries than it does within the EU-15 Member States. For example, in 1997, the average percentage of the population employed in agriculture within the European Community was 5 per cent, whilst that of the ten central and eastern European Member States was 21 per cent.[147] Similarly, whilst agriculture accounted for an average 1.7 per cent of gross domestic product within the economies of the EU-15 Member States, the equivalent average figure for the ten new central and eastern European Member States was 6.8 per cent.[148] Many, though by no means all, of the new central and eastern European Member States have large farming sectors of central importance to their economies. Additionally, in most cases, these farming communities are composed of large numbers of small

144   See generally, Van den Bossche, P. (2005), *The Law and Policy of the World Trade Organisation* (Cambridge: Cambridge University Press).

145   Legg, W. The Environmental Effects of Reforming Agricultural Policies, in Brouer, F., Lowe, P. (eds) (2000), *CAP Regimes and the European Countryside* (Wallingford: CAB International) 21.

Equally, to put these subsidy levels into perspective, Legg also reported that Iceland and Japan had PSEs of over 60 per cent and that Switzerland and Norway had PSEs of over 70 per cent.

146   These countries are the Czech Republic, Estonia, Hungary, Poland, Slovenia, Bulgaria, Romania, Slovakia, Latvia and Lithuania.

147   European Commission (1997), *The Agricultural Situation in the European Union 1998 Report* (Luxembourg: Office for Official Publications of the European Communities) T24.

148   *Ibid.*

farms.[149] The European Commission noted that, without reform, the extension of the Common Agricultural Policy to these new Member States would have created a major budgetary expense for the Community and also made it difficult for the Community to meet its commitments under the GATT Agriculture Agreement.[150]

A final economic force for reform of the Common Agricultural Policy was the fact that agriculture was no longer the predominant employer in rural areas. Agricultural mechanisation, together with the encouragement to amalgamate and enlarge smaller farms into larger more viable units, has been accompanied by reductions in agricultural employment. For example, in 1968, agriculture accounted for 12 per cent of total employment across the European Community.[151] In 1998, however, the European Commission noted that,[152]

> In terms of regional income and employment, agriculture no longer forms the main base of the rural economy. It represents only 5.5 per cent of total employment on average and in very few regions is its share higher than 20 per cent. The long term trend is a further drop in the numbers of farmers, at a rate of 2–3 per cent per year.

In this situation, as previously noted in Chapter 1, the Community recognised a need to develop the Common Agricultural Policy into a more broadly based policy providing for the development of rural areas as a whole.[153]

To some extent, it might be argued that these additional reform pressures actually complemented the need to achieve greater environmental protection. For example, the aim of environmental improvement could be said to coincide with the need to reduce agricultural production levels in order to limit agricultural surpluses and European Community budgetary expenditure.[154] Similarly, the prospect

---

149   Perhaps the largest potential problem in this regard was posed by Poland where 20 per cent of the population worked in agriculture, whilst the average farm size was only 7 hectares. See OECD (1995), *Agricultural Policies, Markets and Trade in the Central and Eastern European Countries* (Paris: OECD) 65.

150   European Commission (1997), *Agenda 2000: The Effects of the Union's Policies of Enlargement to the Applicant Countries of Central and Eastern Europe* (Luxembourg: Official Publications of the European Communities) 6.

151   See European Commission (1976), *The Agricultural Situation in the Community: 1975 Report*, (Luxembourg: Office for Official Publications of the European Communities) 140.

152   European Commission (1997), *Agenda 2000: For a Stronger and Wider Union* (Luxembourg: Office for Official Publications of the European Communities) 26.

153   For example, the Cork Declaration, a document created by a European Community conference on rural development, which was held in Cork in November 1996, called upon the European Community to amend the Common Agricultural Policy in order to adopt such an approach.

154   See, for example, Baldock, D. and Lowe, P. The Development of European Agri-Environmental Policy, in Whitby, M. (ed.) (1996), *The European Environment and CAP*

of introducing environmental payments offered the European Community a simultaneous opportunity to pursue the social objectives of supporting family farms and encouraging farmers to remain in farming.[155] Although such considerations may have helped to build the case for the introduction of environmental measures, they also ensured that these measures did not serve solely environmental goals. This, therefore, created the possibility that the social goals pursued by such measures could result in reduced environmental benefits.

Community Awareness of the Environmental Problems Associated with Agriculture

The European Community's growing awareness of the environmental problems associated with agriculture can be seen through the provisions that it made in its Environmental Action programmes. These action programmes describe proposals for future environmental legislation and also map out the future direction of European Community environmental policy. Although they do not have direct legal effect, they clarify the European Community's future environmental objectives. The first environmental action programme was adopted by the Council of Ministers in 1973 and covered the period 1973 to 1976.[156] Today, European Community environmental policy is guided by the sixth environmental action programme, which applies from 2000 to 2010.[157]

The European Community's initial environmental action programme accepted that agriculture had an impact upon the natural environment. However, both it and the second environmental action programme show that the European Community, initially, had only limited knowledge of this environmental impact. Both action plans committed the European Community to conducting scientific studies to examine the nature and scope of the environmental problems associated with agriculture. For example, the first environmental action programme provided for the Community to conduct research on the environmental effects of a range of agricultural activities – single crop farming, intensive fertiliser use, excessive pesticide use, the use of slurry by intensive livestock farms and the environmental impact of land improvement schemes. Similarly, the second environmental action programme called for research to be conducted into the impact that the use of slurry

Reform: Policies and Prospects for Conservation (Wallingford: CAB International) 12.

155   Scheele. M. The Agri-Environmental Measures in the Context of CAP Reform, in Whitby, M., (ed.) (1996), *The European Environment and CAP Reform: Policies and Prospects for Conservation*, (Wallingford: CAB International) 4.

156   [1973] OJ C112/73.

157   Decision 1600/2002/EC of the European Parliament and of the Council laying down the Sixth Environmental Action Programme [2002] OJ L242/1.

had on human health and the environment.[158] Indeed, the need for further research is also evident in the third environmental action programme, which called for the Commission to conduct research on ways to prevent eutrophication.[159] In total the first and second and third programmes covered the period 1973 to 1986. During this period the European Community was on a learning curve. Little substantive commitment to specific action might have been expected, whilst this happened. Some measures were, however, proposed. For example, the second environmental action programme made a commitment to address the environmental problems associated with pesticide use and with intensive livestock farming. In the case of pesticides, action was promised to regulate the marketing of pesticides, to develop alternative methods of pest control and to monitor pesticide residue levels in food.[160] In relation to intensive livestock farming, the Commission was asked to submit legislative proposals on the collection and storage of agricultural waste and on limiting the amount of animal waste that could be spread upon cultivated land.[161] Similarly, the third environmental action programme called for the Commission to scrutinise the potential environmental impact of applications for farm development grants.[162]

During the life of the third environmental action programme, the European Commission conducted a comprehensive review of agricultural policy. In *Perspectives for the Common Agricultural Policy*, the Commission examined the environmental problems associated with agriculture and also the socio-economic problems that were being experienced by the industry.[163] The Commission recognised that agriculture had been identified as:[164]

> a cause – and sometimes even as the major cause – of the extinction of species of flora and fauna and of the destruction of valuable ecosystems such as wetlands, and in some cases have increased risks of ground and surface water pollution.

The Commission's green paper recommended that agriculture should be treated in the same way as other sectors of the economy and made subject to 'reasonable public prescriptions' and controls designed to avoid the deterioration of the environment.[165] It recommended that common action should be taken to control

---

158   See [1977] OJ C139/1 at 20–22. The second environmental action programme covered the period 1977 to 1981.

159   See [1983] OJ C46/1 at 10. The third environmental action programme covered the period 1982–1986.

160   [1977] OJ C139 at 21.

161   *Ibid.* 22.

162   [1983] C46/12. This was part of a wider examination of the environmental impact of Community structural policies as a whole, encompassing also European Community financial assistance to regional, industrial energy, transport and tourism policies.

163   COM(1985) 333/2 final.

164   *Ibid.* 50.

165   *Ibid.*

the environmental problems associated with intensive livestock production. In particular, that planning restrictions should apply when farm buildings were to be constructed for this purpose.[166] The Commission also recommended that major land use changes, such as the re-parcelling of land, the development of farm roads or the drainage of agricultural land should also become subject to planning permission and that farmers should be required to undertake an environmental impact assessment as part of that planning process.[167] It also suggested that measures be taken to encourage environmentally friendly farm management practices – such as the development of buffer zones where agricultural land lay next to a nature reserve or where such zones would protect surface or ground waters from pollution.[168] In areas in which the environment was particularly threatened, the Commission proposed that such environmental farm management measures should be made compulsory.[169] In other areas it was envisaged that these measures would apply on a voluntary basis.[170] The Commission also foresaw that farmers would receive financial payments in return for providing this environmental management.[171] Perhaps most controversially, the European Commission recommended that up to 10 per cent of the European Community's agricultural land should be purchased or rented by public authorities and used, specifically, for environmental purposes such as nature conservation and the development of recreational amenities.[172]

The European Commission's green paper shows that, by 1986, much more was known about the environment effects of modern agriculture. This also coincided with the insertion of the environmental title into the European Community Treaty, by the Single European Act, which included the assertion that environmental protection should become an integral part of all Community policies.[173] Community awareness of the environmental issues associated with agriculture also became evident when the European Community adopted its fourth environmental action programme in 1987.[174] Indeed, this plan set the specific goal of encouraging agricultural practices that were environmentally beneficial.[175] During the life of the fourth environmental action programme, the European Commission published a

---

166   *Ibid.* 51.

167   *Ibid.*

168   *Ibid.* 52. Within such zones farm management practices that would be applied would include the suspension of agricultural activity at certain times of the year, limitations upon the use of fertilisers and pesticides, restrictions upon grazing and restrictions upon land use changes.

169   *Ibid.*

170   *Ibid.*

171   *Ibid.*

172   *Ibid.*

173   Found today in the obligation to integrate environmental protection requirements into 'the definition and implementation' of other Community policies is now set out in Article 6 EC.

174   [1987] OJ C328/1.

175   *Ibid.* 3.

further evaluation of the relationship between agriculture and the environment.[176] This re-endorsed the views expressed in the previous green paper. It also pointed to the need to tackle agriculture's role in causing nitrate pollution in surface and ground waters and promoted the development of organic farming and the idea that codes of agricultural good practice should be introduced and provided to every farmer.[177]

By 1991, when the European Community began to develop proposals to reform the Common Agricultural Policy, the role that the agricultural support system had played in encouraging agricultural intensification and environmental damage was well known. In putting forward proposals for reform, the European Commission noted that farmers had a dual function – producing food and protecting the environment:[178]

> The activity of producing has traditionally been focused on producing food. While this will remain the primary focus of production, growing emphasis must be put on supplying raw materials for non food uses. Concern for the environment means that we should support the farmer also as an environment manager through the use of less intensive techniques and the implementation of environment-friendly measures.

Similarly, in 1999, when the European Community implemented a second wave of reforms to the Common Agricultural Policy, the European Commission suggested that:[179]

> The philosophy underpinning the environmental aspects of the CAP reform is that farmers should be expected to observe basic environmental standards without compensation. However, wherever society desires that farmers should deliver an environmental service beyond this base-line level, this service should be specifically purchased through agri-environmental measures.

Today, the twin track approach advocated by the European Commission forms the core of the Community's environmental policy in relation to agriculture. It is clear, therefore, that environmental issues represent both a problem to be tackled and also an opportunity to support farmers who undertake measures to enhance the environmental credentials of their farms. This approach is evident in recent environmental action programmes and other Community environmental policy documents. In 1993, the Community's fifth environmental action programme

---

176   European Commission, *Environment and Agriculture*, COM (1988) 338 Final.

177   *Ibid.* 15.

178   European Commission, *The Development and Future of the CAP: Reflections Paper of the Commission*, COM (1991) 100 Final, 10.

179   European Commission, *Directions Towards Sustainable Agriculture*, COM (1999) 22 final, 28.

identified agriculture as one of five economic sectors that had major impacts upon the environment.[180] Currently, the sixth environmental action programme highlights the role that agriculture can play in protecting biodiversity and landscapes. In particular, the current action plan reflects agriculture's dual role. It calls for greater steps to be taken to integrate environmental protection measures into agricultural production policies, whilst also seeking to ensure that greater financial resources are available to support farming practices that protect and enhance the environment.[181] Elsewhere, this strategy is also evident in other European Community environmental policy documents.[182] Additionally, the Commission's proposals for the 2003 Mid Term Review of the Common Agricultural Policy, and 2008 Health Check continued to place these goals at the centre of the reform process.[183]

180   *Towards Sustainability: A European Community Programme of Policy and Action in Relation to the Environment and Sustainable Development*, [1993] OJ C138/1, 28.

181   [2002] OJ L242/1, Articles 6(2) (f) and 4(10) respectively.

182   See, for example, the European Commission's proposed strategy for sustainable development, which was endorsed by the European Council at its 2001 Gothenburg summit: European Commission, *A Sustainable Europe for a Better World: A European Union Strategy for Sustainable Development*, COM(2001) 264, the European Commission's Biodiversity Action Plan for Agriculture: European Commission, *Biodiversity Action Plan for Agriculture*, COM(2001) 162 final and the Commission's proposals for halting biodiversity loss by 2010: European Commission, *Halting the Loss of Biodiversity by 2010 – And Beyond*, COM (2006) 216 final.

183   See European Commission (2002), *Mid-Term Review of the Common Agricultural Policy*, COM (2002) 394 and European Commission (2007), *Preparing for the 'Health Check' of the CAP Reform*, COM (2007)722.

Chapter 3
Integrating Environmental Protection into Agricultural Production Policies

Reforming Agricultural Production Policy

Chapter 2 highlighted the role that the European Community's agricultural production policies have had in encouraging widespread environmental damage. It should also be recognised that these policies have undergone radical reforms over the last two decades. These reforms have often been prompted by economic, rather than environmental, factors. However, each reform has directly influenced production decisions taken by individual farmers and, consequently, has had an important influence upon the rural environment.

Piecemeal Production Policy Reforms in the 1970s and 1980s

Production policy reforms began, essentially, in the late 1970s and early 1980s. During this period a number of incremental changes were introduced in individual sectors. These reforms were largely motivated by the Community's desire to limit the impact that surplus production was having upon the Community budget. At the present time, three commodity groups: arable crops, dairy produce and beef production account for 60 per cent of the Community expenditure on agricultural production.[1] It is, perhaps, not surprising that these commodities were at the forefront of the Community's early reform measures. A variety of methods were used to try to discourage increased production. For example, co-responsibility levies were introduced for milk products in 1977[2] and for arable farmers in 1986.[3] These made small reductions in prices received by farmers, with the Community then using the monies saved to help fund continued purchasing of surplus produce or the payment of export refunds when that produce was exported to non Member States. The Community also made use of guarantee thresholds. For example, a guarantee threshold was introduced for arable crops in 1988. This limited the amount of surplus produce that could be bought by intervention agencies at the established

---

1 See European Commission (2005), *The Agricultural Situation in the European Union, 2004 Report* (Luxembourg: Office for Official Publications of the European Communities).

2 Council Regulation 1079/77, [1977] OJ L131/6.

3 Council Regulation 1579/86, [1986] OJ L139/29.

intervention price.[4] Linked to this threshold, an additional co-responsibility levy was made at the beginning of each marketing year, which would then be refunded at the end of that year if the guarantee threshold had not been exceeded. Similarly, in 1982, a guarantee threshold was also introduced for dairy farmers.[5] On this occasion, however, exceeding the threshold resulted in the introduction of yet another device – the production quota.

Milk quotas are, perhaps, the best known example of measures introduced to discourage increased production. The milk quotas were designed to remove any financial incentive for farmers to increase their production levels beyond their quota allocation. They were introduced by Regulations 856/84[6] and 857/84.[7] The initial milk quotas were set, in 1981, based on farmers' production levels plus 1 per cent.[8] They were backed up by a system of levies intended to heavily penalise production levels that exceeded allocated quota.[9] Milk quotas still remain in place today, but are due to be abolished in 2015.[10]

One other method of discouraging increased production was to limit the availability of the intervention purchasing system. As explained in Chapter 1, the intervention system was based on a system of intervention prices, for particular commodities, that were fixed by the European Community. The system was intended as a floor for Community market prices. It also provided a method through which intervention agencies could purchase surplus produce, at the intervention price, in commodities such as arable crops, beef and dairy produce. Reforms to the intervention system during the 1980s made it more difficult or less viable for farmers to rely upon intervention purchasing as a market for unsold produce. The measures introduced took various forms. Perhaps the simplest was to increase the quality requirements that were demanded of produce offered for sale to national intervention agencies.[11] Alternatively, restrictions were placed

---

4   Council Regulation 1079/88 [1988] OJ L110/7, which set the guarantee threshold at 160 million tonnes.

5   Council Regulation 1183/82 [1992] OJ L140/1.

6   Council Regulation 856/84 [1984] OJ L90/10.

7   Council Regulation 857/84 [1984] OJ L90/13.

8   Subsequently these quota levels were reduced by subsequent legislation, in an effort to reduce the levels of Community milk production. These reductions were made under Council Regulation 1335/86 ([1986] OJ L119/19), Council Regulation 775/87 ([1987] OJ L78/5) and Council Regulation 1109/88 ([1988] OJ L110/27).

9   Initially, Council Regulation 1305/85 ([1985] OJ L137/12) fixed these levies at 75 per cent of the target price of milk for milk sales from farms and 100 per cent of that target price for sales to dairies. By 1992, under Council Regulation 3950/92 ([1992] OJ L137/12) this had been raised to 115 per cent in both cases.

10   See, European Commission, *Proposal for a Council Regulation establishing common rules for direct support schemes for farmers under the common agricultural policy and establishing certain support schemes for farmers*, COM(2008) 306 (final) 9.

11   See, for example Usher, J.A. (1988), *Legal Aspects of Agriculture in the European Community* (Oxford: Clarendon Press Ltd) 62. He notes there that this method was adopted

upon the times of year during which intervention purchasing would be available. For example, under Regulation 773/87 intervention agencies would only be required to purchase skimmed milk powder between 1st March and 31st August each year.[12] Similarly, for arable farmers, Regulation 1900/87 established an intervention period of 1st October to 31st May for cereal crops.[13] One other reform saw ceilings being introduced to limit the amount of particular produce that could be bought into intervention. When the ceiling was passed, intervention spending could be suspended, as it was the case of skimmed milk powder.[14] Alternatively, passing the intervention ceiling might simply result in greater restrictions being placed upon the availability of further intervention purchasing. For example, an intervention ceiling of 180,000 tonnes was adopted for butter. Once this ceiling was exceeded then further intervention purchasing of butter could only occur if average Community market prices dropped to 92 per cent of the intervention price for butter and to 90 per cent of that price should intervention stocks reach 250,000 tonnes.[15] Elsewhere, another reform to the intervention system simply lowered the prices at which intervention purchasing would be triggered. For example, for beef farmers, Regulation 467/87 created a situation in which the fact that market prices had fallen to the level of the intervention price established by the Community no longer triggered intervention purchasing of beef.[16] Instead, such purchasing could only occur when average Community market prices were less than 91 per cent of Community's intervention price and the average regional or national market price for the farmers' particular area was less than 87 per cent of that intervention price.

Collectively, these reforms show that the Community was determined to put the brake upon increased agricultural expansion. However, these measures were motivated purely by economic considerations. Discouraging individual farmers from intensifying their production did also, potentially, create beneficial side-effects for the environment. But, the environmental benefits from these policies were purely incidental to the broader economic benefits that Community policy-makers were, in reality, more concerned with.

---

in relation to cereals, when Council Regulation 1580/86 ([1986] OJ L139/34) raised the quality standards required of wheat and other cereals that were offered for sale to intervention agencies.

12   Council Regulation 773/87 [1987] OJ L78/1.

13   Council Regulation 1900/87 [1987] OJ L182/40, the Regulation also established an amended intervention period for southern European Member States. For Greece, Italy, Portugal and Spain this ran from 1st August to 31st May in each marketing year.

14   Council Regulation 1112/88 [1988] OJ L110/32. This applied in relation to an intervention ceiling of 100,000 tonnes.

15   *Ibid.*

16   Council Regulation 467/87 [1987] OJ L48/1.

*Production Policy Reforms in the 1990s: The MacSharry and Agenda 2000
Reform Packages*

On 30 June 1992 the Community adopted a reform package known as the
MacSharry reforms.[17] This was followed by the Agenda 2000 reforms adopted
on 17 May 1999.[18] Each of these packages introduced a range of reforms across
a number of common organisations. Once again, the common organisations in
cereal crops, dairy and beef produce were at the forefront of these reforms.[19]

In essence, both reform packages followed the same broad agenda: they sought
to increase the market orientation of the Community's agricultural production
policy and to liberalise agricultural trade. In doing so, they brought Community
produce prices closer to the lower prices prevailing upon world markets. This, for
example, limited the need for the Community to pay export refunds and, therefore,
helped the Community to meet the commitments that it had made in the 1994
GATT Agriculture Agreement.[20]

The MacSharry and Agenda 2000 reforms both introduced lower commodity
prices for arable crops and beef produce. For example, Regulation 1766/92
introduced a 29 per cent reduction in target and intervention prices for arable crops
over a three year period,[21] and Regulation 1253/99 then introduced a further 15 per
cent price reduction in these prices.[22] Similarly, intervention prices paid for beef
were also reduced in both 1992 and 1999.[23] The Agenda 2000 reforms also made
reductions in the fixed intervention prices for butter and skimmed milk[24] and the
Mid Term Review introduced further reductions in these prices.[25]

These reforms also retained some of the earlier measures that sought to
discourage increased production. For example, the use of milk quotas was extended
to, at least, March 2015. Similarly, the availability of intervention purchasing
continued to be restricted. Henceforth, intervention purchasing was to operate as a

---

17   Named after Raymond MacSharry, who was then European Commissioner for
Agriculture.

18   Named after the European Commission proposals, which were designed to prepare
Community policies for the challenges of the new millennium: European Commission,
Agenda 2000: For a Stronger and Wider Union, COM (97) 2000.

19   Though others such as the common organisations concerned with tobacco
production, rice and fruit and vegetables were also affected.

20   See Chapter 2 for further details of these commitments.

21   Council Regulation 1766/92 [1992] OJ L181/21.

22   Council Regulation 1253/99 [1999] OJ L160/18.

23   See Council Regulation 2066/92, [1992] OJ L215/49 and Council Regulation
1254/99, [1999] OJ L160/21.

24   See Council Regulation 1255/99, [1999] OJ L160/48.

25   See Council Regulation 1787/2003, [2003] OJ L270.

safety net, where market prices had experienced sharp falls, rather than providing a market for surplus produce.[26]

At first sight, widespread commodity price reductions and continued restrictions on intervention purchasing might seem to be environmentally beneficial. For example, lower prices might be expected to encourage farmers to make less use of inputs such as artificial fertilisers and pesticides. After all, as one commentator has observed, 'Wheat produced at £80 per tonne simply does not buy as much chemicals and fertiliser... as it does at £110 per tonne.'[27] However, lower prices could also have adverse environmental side-effects. As Dhondt has pointed out, price reductions in one agricultural sector can encourage increased production in others, for example lower cereal prices translate into lower feed prices for intensive livestock producers and, thereby, encourage greater levels of beef production.[28] They could also make conservation-minded farmers less willing to, voluntarily, assume the cost of maintaining wildlife habitats on their land. Price reductions also increase the risk of farm abandonment, with the particular danger that habitats on extensively managed land, of high nature conservation value, could be lost as that land reverts to shrub.[29] On their own, therefore, lower commodity prices and restrictions on intervention purchasing would be likely to produce mixed results for environmental protection. Within the Community, however, farmers have been compensated by the introduction of a number of compensatory payments, such as the arable area payment and premiums for beef and dairy farmers. As commodity prices were steadily reduced, by successive reform packages, the level of these direct payments gradually increased.

More recently, the 2003 Mid Term Review and 2008 Health Check have sought to amalgamate these various individual direct payments into one single farm payment.

The use of these direct payments represented a change in the Community's philosophy concerning farm support. Where previously the Community had sought to protect farm incomes through agricultural produce prices, it was now

---

26   For example, in relation to beef, Council Directive 1254/99, [1999] OJ L160/21, Article 47(3) provided for intervention purchasing to be available where, for 2 consecutive weeks, both the average market price in a Member State or region is less than 80 per cent of the intervention price established by the European Community and the average Community market price also falls below 84 per cent of that intervention price. From 1 July 2002 this was amended so that intervention purchasing of beef became available when the average market price in a Member State or region of a Member State remained below €1560 per tonne for two consecutive weeks.

27   Harvey, D. (1991), *The CAP and Green Agriculture* (London: Institute for Public Policy Research: London).

28   Dhondt, N. (2003), *Integration of Environmental Protection into other EC Policies* (Groningen: Europa Law Publishing) 257.

29   Baldock, D., Beaufoy, G. (1993), *Nature Conservation and New Directions in the EC Common Agricultural Policy* (London: Institute for European Environmental Policy) 117.

doing so by providing direct support for producers themselves. In practice, direct payments had begun to become part of the Common Agricultural Policy as part of the piecemeal reforms introduced during the 1980s. However, following the MacSharry reforms they gained a more central role in production policy.

Direct payments have always been a feature of the common organisation in sheep and goats meat, since that common organisation was first introduced in 1980.[30] Sheep Annual Premium payments initially compensated sheep farmers when the average price for sheep meat in their region fell below a basic price fixed by the Community.[31] In 1985, following Greece's accession to the European Community the payment was extended to include goat meat.[32] More recently, following a reform of the common organisation in sheep and goat meat in 2001, direct payments to sheep and goat farmers were transformed into a fixed payment to all sheep farmers in place of a variable compensatory premium.[33]

In 1987, direct payments were also introduced within the common organisation for beef and veal. These payments compensated farmers for the limitations that had been placed on intervention purchasing and for reductions in the intervention purchasing price.[34] As a consequence of subsequent reforms, a range of direct payments became available to beef farmers. These included special beef premiums, suckler cow premiums, deseasonalisation premiums and slaughter premiums. Special beef premiums could be claimed by farmers on male beef cattle. Farmers could claim the payment once only on up to 90 bulls, as soon as the bulls reached the age of nine months. Farmers could claim the payment twice in the life of up to 90 steers, when each steer reached the age of nine months and again when they reached the age of 21 months.[35] In contrast, suckler cow premiums were annual payments claimed on the number suckler cows retained by beef farmers.[36] Eligibility for suckler cow premiums was subject to a number of conditions. Farmers were required to retain at least 80 per cent of the suckler cows, on which they had claimed the premium, for a minimum of six months after they lodged their claim, and at least 20 per cent

---

30   The common organisation in sheep-meat was introduced by Council Regulation 1837/80, [1980] OJ L183/1.

31   Council Regulation 1837/80, Article 5.

32   Council Regulation 3523/85, [1985] OJ L336/2, the payment was made at 80 per cent the rate of sheep annual premium.

33   The reform was introduced by Council Regulation 2529/2001, [2001] OJ L341/3, Article 4.

34   By Council Regulation 467/87, [1987] OJ L48/1.

35   See, for example, Council Regulation 1254/99 [1999] OJ L160/21, Articles 4(1) and 4(2). Under Article 3 of this Regulation a bull is defined as being an uncastrated male bovine animal, whilst a steer is defined as being a castrated male bovine animal.

36   See Council Regulation 1254/99, Article 6(2). Under Article 3 of the Regulations a suckler cow is defined as being a cow which belongs to a meat bread or is born of a cross with a milk breed, and belongs to a herd intended for rearing calves for meat production.

of heifers upon which the payment was claimed.[37] Additionally, it was intended that the payments should support farmers specialising in beef farming, therefore limitations were placed upon farmers' involvement in dairy farming. Member States were given the choice of either setting the condition that eligible farmers must not have supplied any milk or milk products from their farms (excluding direct sales to consumers from the farm) or that the milk supplied did not exceed an annual quota of 120,000 kilograms.[38] A slaughter premium was also paid on the slaughter or export from the EU of eligible cattle.[39] In certain conditions a separate deseasonalisation premium was also payable on the slaughter of steers.[40] Finally, the Agenda 2000 reforms also introduced the concept of a 'national envelope', into the common organisations for beef and veal. Member States were each allocated a maximum sum of money and given the discretion to allocate that money to beef farmers on either a headage or an area basis.[41]

For arable farmers, Regulation 1765/92 introduced the concept of the arable area payment.[42] These payments were increased in value under Regulation 1251/99.[43] They were payable for land on which arable crops were being grown or which was subject to compulsory set aside requirements.[44] The Community also put in place measures to ensure that farmers could not simply farm for the arable area payment alone, for example by sowing a thin layer of seed and not harvesting the crop.[45] Claimants were required to ensure that their crops had been

37   *Ibid.* Under Article 3 a heifer is defined as being a female bovine animal from the age of eight months which has not yet calved.

38   *Ibid.* Article 6(2) also allowed Member States, on the basis of objective criteria, to change or waive this 120,000 kg limit.

39   Council Regulation 1254/99, Article 11. Eligible animals were either (a) bulls, steers, cows and heifers from the age of eight months and (b) calves of more than one and less than seven months old which had a carcass weight of less than 160 kg.

40   Council Regulation 1254/99, Article 5. Deseasonalisation premium would be payable where the number of steers slaughtered in a given year exceeded 60 per cent of total annual slaughterings of male bovine animals and the number of steers slaughtered during the period 1st September to 30th November in a given year exceeds 35 per cent of total annual slaughterings.

41   Council Regulation 1254/99, Article 14 and subject to various conditions set out in Articles 15–19.

42   Council Regulation 1765/92 [1992] OJ L181/12.

43   Council Regulation 1251/99 [1999] OJ L160/1.

44   Council Regulation 1251/99, Article 7. This also required that the land had not been under permanent pasture, permanent crops, forest or in non agricultural use on 31 December 1991. Under Article 3 and the annex to the Regulation 'permanent pasture' is defined as being non-rotational land used for grass production (sown or natural) on a permanent basis. Permanent is defined as being for five years or more. 'Permanent crops' are defined as being non-rotational crops that occupy the ground for five years or longer and yield repeated harvests.

45   Neville, W., Mourdant, F. (1993), *Guide to the Reformed Common Agricultural Policy* (London: Estates Gazette Ltd) 22.

sown in accordance with locally recognised standards and were maintained, at least until the beginning of flowering, in normal growth conditions.[46] The arable area payment was calculated on the basis of the average yield per hectare for each arable region.[47] This was based on the yields produced in the five-year period from the marketing year 1986/87 to the marketing year 1990/91, with the the highest and lowest yields being excluded.[48] The arable area payment for each farmer was then calculated by multiplying the average yield per hectare by a basic payment per tonne fixed by the Community and multiplying the figure obtained through this calculation by the number of hectares of eligible land farmed by each claimant.[49] The average yield per hectare was also used to identify those farmers who had an obligation to set land aside from production. Based on the average yield per hectare, farmers who claimed arable area payments on an area greater than that required to produce 92 tonnes of arable crops were required to set aside, from production, a pre-determined percentage of their arable.[50] Alternatively, they could opt to use this land to grow crops that would be used within the Community to manufacture products not intended for human or animal consumption. Farmers who did not meet this 92 tonne threshold could avoid this compulsory set aside requirement all together.[51]

Dairy farmers have also been compensated for reductions in the intervention prices fixed for butter and skimmed milk by the introduction, in 2005, of a dairy premium.[52] In this case, the premium was to be calculated by multiplying each dairy farmer's annual production quota, in tonnes, by an amount fixed by the Community.[53] Member States were again allocated an additional 'national envelope', which they were authorised to allocate to dairy farmers as either a supplement to dairy premium and/or as an area payment.

Agricultural Production Policy Today: The Impact of the 2003 Mid Term Review and of the 2008 Health Check

The MacSharry and Agenda 2000 reforms linked the direct payments made to farmers with their agricultural production. This link is often referred to as the 'coupling' of direct payments to agricultural production requirements. It had the

---

46   Commission Regulation 2780/92, [1992] OJ L281/5, Article 5.
47   Council Regulation 1251/99, Article 3. This also required Member States to identify to the European Commission each of the regions within their territory.
48   *Ibid.* Article 3(5).
49   *Ibid.* Article 4(1). The 'basic amount per tonne' was fixed by this article as being €58.67 for the marketing year 2000/2001 and €63 in subsequent years.
50   Council Regulation 1765/92, Article 7.
51   *Ibid.*
52   Introduced by Council Regulation 1255/99, [1999] OJ L160/73.
53   *Ibid.* Article 16. The amount fixed by the Community was €5.75 in 2005, rising to €11.49 in 2006 and €17.24 in 2007 and subsequent years.

consequence of tying farmers to particular types of agricultural production. In order to remain eligible to receive direct payments from one sector, farmers were reluctant to diversify into other sectors.

In 2002, in putting forward its initial proposals for the Mid Term Review reforms, the European Commission sought to remove any coupling of agricultural production with direct payments.[54] They envisaged that the various direct payments made through individual common organisations would be replaced by one single farm payment. This single farm payment would then be calculated on the basis of past production levels and would not require farmers to continue farming in any particular sector. The Commission noted that this move toward a fully decoupled payment system would complete the shift in the European Community's support policy, from product to producer.[55] Additionally, the European Commission suggested that entitlements to single farm payments should be calculated solely on the basis of past production. They argued that this would remove any incentive for farmers to increase production levels and thereby obtain higher payments and further help to integrate environmental protection considerations into agricultural production policy.[56]

Ultimately, under Regulation 1782/2003, the single farm payment scheme was introduced in January 2005.[57] Though, 'where specific agricultural conditions warrant', Member States were authorised to delay the introduction of the single farm payment until 1st January in either 2006 or 2007.[58] The single farm payment scheme drew together the various arable and livestock payments into one consolidated payment. It was to be calculated on the basis of the average payments that individual farmers had received under the pre-existing regimes in the calendar years 2000–2002.[59] Each Member State was also required to ensure that their total annual expenditure on the single farm payment did not exceed a national ceiling established by the Community.[60] Member States were required to decide, by 1st August 2004, whether to apply the single farm payment scheme on a regional or national basis.[61] Member States that chose a regional approach were then required to divide their national ceiling into separate regional ceilings.[62] A simplified version of the single farm payment, 'the single area payment', was also introduced for new Member States who joined the European Community in either 2004 or 2007. This provided uniform payments per hectare of eligible land,

54   European Commission, *Mid Term Review for Sustainable Agriculture*, COM (2002) 394.

55   *Ibid.* 19.

56   *Ibid.*

57   Council Regulation 1782/2003 [2003] OJ L270, Articles 33 and 156.

58   *Ibid.* Article 71.

59   *Ibid.* Article 38.

60   *Ibid.* Article 41.

61   *Ibid.* Article 58.

62   *Ibid.* Article 58(3).

subject again to fixed national spending ceilings. The single area payment was designed to be an interim measure that would enable the farming sector in each new Member State to adjust to the Common Agricultural Policy, before migrating to the single farm payment.

The Single Farm Payment scheme adopted by the Community contained several important differences from the one initially proposed by the European Commission. One difference was that, instead of simply basing all payments upon farmer's actual historic production levels, Member States were also given the option of calculating payments on an area basis. Under this option, payments could be calculated by dividing the monies available to a Member State or region amongst eligible farmers on the basis of the area of land that they farmed.[63] Alternatively, a variation of this area based approach, authorised the payment levels to be flexible according to the number of hectares that, in 2003, each farmer had under grassland, permanent pasture or other agricultural use.[64]

Perhaps an even more important change from the initial Commission proposals was that the Single Farm Payment scheme also enabled national agricultural authorities to operate a hybrid scheme, alongside slimmed down versions of pre-existing direct payments, such as arable area payments or livestock headage payments. This arose because some Member States feared that a fully decoupled regime could create risks such as abandonment of production and social or environmental problems in areas where there were few other economic options open to farmers.[65] Where this option was chosen, the amounts paid to each farmer under the Single Farm Payment would be reduced. The remaining portion of the direct payments would then be paid as arable area payments or livestock headage payments. A range of options existed.[66] For example, up to 25 per cent of the portion of any regional or national ceiling under the Single Payment Scheme which related to arable area payments could continue to be paid as arable area payments. Similarly, up to 50 per cent of the amount allocated for sheep farming could be retained as a headage payment on the number of sheep owned. Equally, in the case of beef farmers, up to 100 per cent of suckler cow premium and up to 75 per cent of beef special premium could be retained within a particular Member State or region. The fact that these livestock based payments were based upon current livestock production levels, rather than historic production as the Single Payment Scheme had envisaged, created the potential to undermine the Commission's environmental objectives. Where adopted, they provided a financial incentive for farmers to maintain livestock numbers at a level that maximised the direct payments that they received. Once again, the fact that quotas attached to

---

63   *Ibid.* Article 59.

64   *Ibid.* Article 61.

65   European Commission, Proposal for a Council Regulation establishing common rules for direct support schemes for farmers under the common agricultural policy and establishing certain support schemes for farmers, COM(2008) 306 final 5.

66   *Ibid.* Council Regulation 1782/2003, footnote 57 above, Article 64.

these payments took no account of the environmental capacity of individual farms meant that potential for environmental damage continued to exist. It can be argued that these measures were also accompanied by the introduction of compulsory cross-compliance, under which all recipients were required to meet environmental standards. These cross-compliance measures are examined below. They include an obligation for farmers to avoid causing environmental damage through overstocking. However, welcome though these cross-compliance measures may be, they also serve to illustrate the ability of one part of agricultural policy to work directly against another. On the one hand, by allowing Member States to retain livestock payments that are based upon existing livestock levels the Community preserved the incentive for individual farmers to maximise their livestock levels and, potentially, to cause environmental damage through overgrazing. On the other hand, as examined below, prevention of over-grazing, is one of the goals of the Community's cross-compliance policy. This seemed a rather strange way to set out to achieve the Community's overarching goal of integrating environmental protection requirements into agricultural production policy. This anomaly was addressed in the 2008 Health Check reforms.[67] The European Commission proposed that discretion granted for Member States, to retain production based payments, should be reduced. Even greater emphasis would be placed upon the single farm payment scheme, which would continue to be calculated on the basis of historic production levels. The only exceptions to this were in relation to suckler cow, sheep and goat meat premiums.[68] The European Commission believed that this was necessary to allow Member States to continue to provide these payments on the basis of current livestock numbers so that adequate support could continue to be available to farmers based in areas in which few other economic activities were available.

The 2008 Health Check also placed a floor on direct payments. In recognition of the fact that almost 50 per cent of farmers receiving direct payments in the EU-25 received less than €500 per annum, the Health Check introduced a requirement that, from 2010, payments should only be made to those who were either eligible to receive at least €100 a year or farmed a minimum eligible area of at least 1 hectare.[69]

Finally, the Mid Term Review also authorised each Member State to retain up to 10 per cent of the national ceiling allocated to it by the Single Farm Payment scheme.[70] The monies retained operate, effectively, as a national envelope. Member States are entitled to use this money to make additional payments to encourage particular types of farming that are important for the protection or enhancement of

---

67   European Commission, *Proposals for a Council Regulation establishing common rules for direct support schemes for farmers under the common agricultural policy and establishing certain support schemes for farmers*, COM(2008) 306 final, 5.

68   See Council Regulation 73/2009, [2009] OJ L30/16, Articles 52 and 53.

69   *Ibid.* Article 28.

70   Council Regulation 1782/2003, footnote 57 above, Article 69.

the environment or for improving the quality or marketing of agricultural products.[71] The 2008 Health Check provided for this money to become available for a broader range of measures.[72] These include using the monies to make payments to farmers just commencing farming and existing farmers in agriculturally disadvantaged areas or areas subject to restructuring or farm development programmes.

Using Direct Payments to Discourage Intensification

The Community's change from price support to producer support offered several potential environmental benefits. Firstly, it provided the opportunity to attach environmental conditions to direct payments made to farmers. This will be examined below, in the section on cross-compliance. Secondly, the Community could continue to discourage intensification, by placing limits upon the availability of direct payments. This applied where direct payments were coupled to agricultural production. When the Single Farm Payment was, subsequently, introduced, these restrictions no longer applied since the payments would now be based upon historic production levels. However, they did continue to apply in Member States that chose to implement a hybrid version of the Single Farm Payment, since part of their payments continued to be linked to production levels.

One example, of a limitation upon the availability of direct payments, arose in relation to the payments made to dairy farmers. These were based upon the milk quota actually allocated to each farmer.[73] This quota, together with the levy applied by the Community for increased production, discouraged farmers from producing beyond this level. In addition to dairy farming, quotas also applied to the payments made to beef farmers and provided similar discouragement. At Member State level, national ceilings were set for beef special premium and suckler cow premium. For beef special premium this ceiling determined the number of cattle upon which this premium could be paid in full. If the ceiling were exceeded in any Member State, payments to all participating farmers in that State would be proportionately reduced.[74] In contrast, farmers claiming suckler cow premium were each given a quota, based upon the number of suckler cows that they owned in December 1999.[75]A national ceiling also established the maximum number of cattle for which suckler cow premium could be paid in each Member State.[76] If necessary, Member States would then have to reduce the quotas allocated to each

71   Council Regulation 1782/2003.

72   European Commission, footnote 67 above, 6 and Council Regulation 73/2009, footnote 68 above, Article 41.

73   *Ibid.*

74   Council Regulation 1254/99, [1999] OJ L160/21, Article 4 (4) and Annex I.

75   *Ibid.* Article 7.

76   *Ibid.* Article 7(2) and Annex II.

individual farmer so that the total number of cows that the premium was claimed on would not exceed the national ceiling.[77]

Claimants for both beef special premium and suckler cow premium also had to comply with stock density limits. These were calculated in, notional, 'livestock units'[78] and based upon the forage area of their farms.[79] Initially, the stock density limits were set at 3.5 livestock units per forage hectare in 1993, falling to 2 livestock units per forage hectare from 1996.[80] More recently, this was further reduced, to 1.9 livestock units from 2002 and 1.8 from 2003.[81]

However, the reality remains that each of these quotas and stock density limits was motivated primarily by economic considerations. They were mechanisms by which the Community could limit its budgetary exposure, since it was the Community budget that would ultimately have to fund these direct payments. In contrast, no account was taken of the environmental capacity of particular farms. Milk quotas and suckler cow premium quotas may have been set for every farm. No consideration was given to the environmental implications of each quota. In contrast, the stocking density limits upon beef premiums were set centrally by the Community and applied uniformly to all farms. In practice, the maximum permitted livestock density per hectare figure was too high to produce widespread environmental benefits. Many beef farmers already had stock densities that were below this limit.[82] Consequently, in many cases, it did not provide any effective incentive against further intensification. In reality, a centrally established stocking density limit simply could not take account of the varying environmental conditions that existed on farms across the European Community.

From an environmental perspective, perhaps the most interesting aspect of beef special premium and suckler cow premium was that farmers became eligible for additional payments if they maintained lower stocking densities. This concept,

---

77   *Ibid.* Article 7(2).

78   For example, Council Regulation 1254/99 Annex III provides for livestock to be attributed the following livestock units each: Dairy cattle, cattle upon which suckler cow premium has been paid, cattle upon which beef special premium has been paid which were aged 2 years or over at the date of claim: 1 livestock unit; cattle upon which beef special premium has been paid which were aged under 2 years at the date of claim: 0.6 livestock units; ewes upon which sheep annual premium has been paid: 0.15 livestock units.

79   Council Regulation 1254/99, Article 12(2) defined 'forage area' as being 'the area of a holding which is available throughout the calendar year for rearing bovine animals, sheep or goats, does not include buildings, woods, ponds or paths, does not include land for which area aid is included under the arable area payment scheme (including set aside land) and does not include land used for permanent crops or horticulture.'

80   Under Council Regulation 2066/92. [1992] OJ L215/49. The maximum stock density requirement did not apply to farms with 15 eligible livestock units or less.

81   By Council Regulation 1512/2001, [2001] OJ L201/1.

82   See Anderson, E., Rutherford, A. and Winter, M. The Beef Regime, in Brouwer, F. and Lowe, P. (eds) (2000), *CAP Regimes and the European Countryside* (Wallingford: CABI Publishing) 55.

the extensification payment, was introduced by the 1992 MacSharry reforms. Originally, farmers could receive an additional 30 ecus per cow, in suckler cow or beef special premium, if they maintained a stocking limit of 1.4 livestock units per forage hectare or less.[83] Additionally, a higher payment could be obtained if the stocking density did not exceed 1 livestock unit per forage hectare.[84] Under Regulation 1254/99 the latter payment was abolished and an increased general extensification payment, of €100 per animal, was made for stocking densities of 1.4 livestock units per forage hectare or less. Alternatively, Regulation 1254/99 also gave Member States discretion to pay smaller extensification premiums to farmers maintaining higher stocking densities. This was intended as a means through which more gradual livestock density reductions could be achieved.[85] The existence of this discretion might, itself, suggest that the Regulation had weak environmental credentials. Though it has been argued that the discretion may have had positive effects on more intensively farmed lands, where higher stocking densities, of between 1.4 and 2.0 livestock units, had traditionally been maintained.[86] Overall, however, the extensification limit of 1.4 livestock units again suffered from the weakness of being set centrally. Once again, no consideration was given to the environmental conditions actually being experienced in various parts of the Community. In some parts of the Community farmers might actually have been causing environmental damage, through overgrazing, by stocking up to 1.4 livestock units per forage hectare.[87] However, these farmers would have been quite entitled to claim extensification premiums.

The environmental credentials of the extensification payment were also undermined by the manner in which it was calculated. When the Community introduced stocking limits and extensification payments in 1992 it did not take all farm livestock into account when implementing the calculation. Instead, only those livestock on which livestock premiums had been paid, along with dairy cattle, were taken into account. Farmers could keep other animals, so that

---

83 Article 4(h) (i) of Council Regulation 806/68 as amended by Council Regulation 2066/92, [1992] OJ L215/49.

84 *Ibid.*

85 Council Regulation 1254/99, Article 13(2) provided that Member States could in the years 2000 and 2001 pay extensfication premiums of €33 per eligible animal to farmers who maintained a stocking density of between 1.6 and 2 livestock units per forage hectare and €66 per eligible animal to farmers who maintained stocking densities below 1.6 livestock units per hectare. In 2002 and subsequent years they were authorised to pay an extensification premium of €40 per eligible animal for stocking densities of 1.4 to 1.8 livestock units per forage hectare and €80 when stocking densities were less than 1.4 livestock units per forage hectare.

86 Anderson, E., Rutherford, A., Winter, M. footnote 82 above at p. 68.

87 For example, Birdlife International, 1997. *A Future for European Rural Environment: Reforming the Common Agricultural Policy* (Brussels: Birdlife International), 35, refers to this precise situation occurring on Spanish dehesas. See also Anderson, E., Rutherford, A., Winter, M. footnote 82 at p. 64.

their actual stocking levels exceeded both the stocking levels required in order to qualify for extensification premiums or indeed suckler cow or special beef premium themselves. Yet those farmers remained eligible for all three payments. Equally, farmers could elect not to claim suckler cow or special beef premium payments upon all their livestock, in order to remain within the 2 livestock per forage hectare limit or to become eligible for extensification payments. The 1999 Agenda 2000 reforms partially addressed this situation.[88] Henceforth, eligibility for extensification payments would be assessed on the basis of the number of male bovine animals, cows and heifers that were actually present on each farm, together with the number of sheep and/or goats for which premium applications had been made.[89] This meant that sheep and goat numbers continued to be calculated on the basis of sheep premium applications, rather than those present on each farm. Overall, the Court of Auditors has concluded that extensification payments actually had little impact upon farmers' production decisions.[90] Most claimants were already practising extensive farming and in some cases could have increased their stocking levels and continued to receive the payment. As such, its principal impact lay in providing an additional source of income for extensive farmers.[91]

The Agenda 2000 reforms also made one further important change to the system of direct payments that existed within production policy. It provided for each Member State to be allocated a fixed amount of Community money, the so called 'national envelope', from which additional payments could be made to beef and, from 2005, dairy farmers.[92] Member States were given broad discretion in deciding how to allocate these monies. They were, however, required to ensure that the extra payments were made on the basis of objective criteria including 'the relevant production structures and conditions, and in such a way as to ensure equal treatment between producers and to avoid market and competition distortions.'[93] Subject to these conditions, Member States had considerable discretion in the manner in which they introduced these additional payments. One option would have been to use these extra payments as a means to provide extra support to farmers engaged in practices supporting the environment. However, the European Commission subsequently noted that, in practice, little attempt was made to use

---

88   See Court of Auditors, Special Report No.5/2002: *Extensification and Payment Schemes in the Common Organisation of the Market for Beef and Veal*, [2002] OJ C290/1, 15.

89   Council Regulation 1254/99, [1999] OJ L160/21, Article 13(2).

90   Court of Auditors, footnote 88 above, 16.

91   See also Cardwell, M. (2004), *The European Model of Agriculture* (Oxford: Oxford University Press) 195.

92   In respect of payments to beef farmers see Council Regulation 1254/99, Articles 14 to 16. For payments to dairy farmers see Council Regulation 1255/99, [1999] OJ L160/48, Articles 17–20.

93   Council Regulation 1254/99, Article 14(1) and Council Regulation 1255/99, Article 17(1).

them to support environmentally friendly farming, the vast majority actually being used to support agricultural production.[94]

Delivering Environmental Protection Through Cross-Compliance

An introduction to Cross-Compliance

In environmental terms, the real significance of the introduction of direct payments into agricultural production policy was not the fact that the Community could use these payments to discourage farmers from increased intensification. It lay in the fact that these payments provided an important opportunity to ensure that farmers fulfilled more specific environmental obligations. The payments could be linked to an obligation that recipients should respect specific environmental conditions. This link forms the basis of the principle of cross-compliance. Proponents of the principle note that it maximises the reach of environmental measures, by ensuring that they are implemented over a much larger area of the countryside than would otherwise be possible under voluntary measures – such as the voluntary environmental land management agreements examined in Chapter 5.[95] The European Community expressed support for cross-compliance measures in its 5th Environmental Action Programme in 1993, noting that the allocation of premiums and other compensatory payments should be 'subject to full compliance with environmental legislation.'[96] Equally, government in the United Kingdom has also been supportive. For example, it previously noted:[97]

> The Government will seek, wherever possible and worthwhile, to develop the integration of agricultural and environmental policies within the European Community, including changes to Community arrangements so that those benefiting from EC support schemes will be required in return to protect and , where possible, enhance the environment on their holdings.

94   European Commission, *Mid Term Review for Sustainable Agriculture*, COM (2002) 394 at 8.

95   See, for example, Potter, C. (1996), *Decoupling by Degrees? Agricultural Liberalisation and its Implications for Nature Conservation in Britain*, Research Paper 196, (Peterborough: English Nature) 13.

96   Council of the European Community, *Fifth Environmental Action Programme in Relation to the Environment and Sustainable Development*. Official Journal [1993] C168/38.

97   United Kingdom Government (1990), *This Common Inheritance: Britain's Environmental Strategy*, Cm.1200 (London: HMSO) 99.

However, the adoption of cross-compliance measures also required majority support in the Council of the Europe. Unfortunately, this level of support was, initially at least, not forthcoming.

Cross-compliance first developed in the United States in the 1970s. The principle initially developed there as an economic measure, rather than environmental one. It was initially developed as a means of ensuring that arable farmers claiming subsidies under different support programmes, for example for both wheat and barley, would have to satisfy the rules of both programmes in order to be eligible for either subsidy.[98] Indeed, within the United States, the term cross-compliance has been used to refer to measures that make such economic linkages between several production programmes. In contrast, the principle, that farmers receiving subsidy payments should meet environmental conditions, has become known as 'resource compliance' or 'conservation compliance'. Consequently, the concept of cross-compliance has developed different meanings in the United States and the European Community. Within the European Community cross-compliance has come to refer to the attachment of environmental conditions to the direct payments made within agricultural production policy.

It was the 1985 Food Security Act, in the United States, that first linked agricultural subsidies to environmental conditions.[99] Three such linkages were created under this Act. First, farmers with arable land in areas identified by the US Government as being seriously threatened by soil erosion were required to implement a conservation plan.[100] Failure to properly implement this conservation plan could result in farmers losing the subsidies payable across their entire farms, not just those payable in relation to the land threatened by soil erosion.[101] Second, a plan entitled 'sodbuster' discouraged farmers from bringing previously uncultivated lands into production within areas threatened by serious soil erosion.[102] Farmers who ploughed up such lands, again, risked losing their entitlement to agricultural subsidies.[103] Third, a programme entitled 'swampbuster' withdrew eligibility for government subsidies from farmers who reclaimed wetlands for use in arable production.[104]

98   Baldock, D. Mitchell, K. (1995), *Cross Compliance within the Common Agricultural Policy: A Review of Options for Landscape and Nature Conservation* (London: Institute for European Environmental Policy) 3.

99   See Farrier, D. Conserving Biodiversity on Private Land: Incentives for Management or Compensation for Lost Expectations? [1995] 19 *Harvard Environmental Law Review* 303.

100   Baldock, D., Mitchell, K. footnote 98 above, 7.

101   *Ibid.*

102   *Ibid.* p. 8.

103   *Ibid.*

104   *Ibid.* p. 8. See also McBeth, D. Wetlands Conservation and Federal Regulation: Analysis of the Food Security Act's Swampbuster Provisions as amended by the Federal Agriculture Improvement and Reform Act of 1996. [1997] 21 *Harvard Environmental Law Review* 201.

In the United States, two main approaches have been identified to environmental cross-compliance: red and green ticket cross-compliance.[105] The programmes introduced by the 1985 Food Security Act are examples of red ticket systems. They enabled agricultural authorities to suspend all or part of the agricultural support payments made to farmers who failed to meet specified environmental conditions. In practice, red ticket schemes could operate in several ways.[106] A minimalist approach would penalise only those farmers who were responsible for significant environmental damage.[107] Alternatively, cross-compliance measures could set stringent conditions that all farmers had to meet. Under this approach, farmers could be required to abide by general obligations – such as the obligation to comply with all Community environmental legislation or with a code of agricultural good practice. They might be required to observe specific conditions, such as minimum and maximum stocking rates or maximum application rates for fertilisers and pesticides.[108] An even more stringent version of red-ticket cross-compliance could combine negative constraints with positive obligations that required farmers to carry out particular works in order to qualify for agricultural payments. This might, for example, include an obligation to create buffer strips between arable fields and watercourses or to maintain hedges and stone walls.

The red ticket approach offered the Community a method by which farmers could be made accountable for the environmental damage that their farming practice was causing. As such, it seemed to provide a direct method to integrate environmental protection considerations into the Common Agricultural Policy, as required under Article 6 of the EC Treaty. Additionally, it also offered the potential to limit budgetary expenditure, since farmers who contravened the principle would not receive Community payments. However, the question remained as to whether Member States would find such constraints upon the farming community to be politically acceptable. A question mark also existed over whether cross-compliance should be used to place positive obligations upon farmers. The danger existed that such an approach would undermine the use of voluntary environmental land management contracts, which many farmers were participating in.[109] Positive measures, required under cross-compliance, could overlap with measures that farmers were being paid to conduct under these management contracts.

---

105   See, for example, Batie, S.S., Sappington, A.G. Cross-Compliance as a Soil Conservation Strategy: A Case Study. [1986] 68 *American Journal of Agricultural Economics* 880.

106   See Baldock, D., Beaufoy, G. (1993), *Nature Conservation and New Directions in the EC Common Agricultural Policy* (London: Institute for European Environmental Policy) 131.

107   *Ibid.* p. 131. The authors note that an informal agriculture Council meeting in 1992 discussed the adoption of such an approach.

108   *Ibid.*, 132.

109   These are examined later, in Chapter 5.

Green ticket cross-compliance is, essentially, the opposite approach to red ticket cross-compliance. In this case, farmers would continue to receive direct payments under production policy, but would also be eligible to receive extra payments in return for meeting particular environmental goals. Again, these environmental conditions could be constructed negatively or positively. However, where a red ticket approach would offer the possibility of budgetary savings, the green ticket approach would require the Community to agree to increased expenditure.

In addition to red and green ticket schemes, a third, orange ticket, approach has also been proposed.[110] This would require farmers to draw up and implement farm conservation plans, as a condition for remaining eligible to receive production based direct payments. Within the Common Agricultural Policy, this approach could require farmers to take part in, what are otherwise voluntary, agri-environmental management programmes in order to qualify for the direct payments available within agricultural production policy. In the United Kingdom, this approach has previously been advocated by the Royal Society for the Protection of Birds.[111] As with the green ticket approach, orange ticket cross-compliance would enable farmers to qualify for both the direct payments made as part of agricultural production policy and also the agri-environmental payments available under the Community's agri-environment policies.[112] This, again, would have had implications for the Community budget, since farmers would be encouraged to enter schemes that cause additional expense for the Community.[113] An orange ticket approach would also have had radical consequences for the future of the Community's voluntary agri-environment measures. Linking these measures to the direct payments made within agricultural production policy would result in a large increase in the numbers of farmers enrolled in them. This might be welcomed as an extension of the environmental reach of these measures. However, many of the newly enrolled farmers would have joined simply out of financial compulsion, to retain access to their direct payments. Consequently, many might be less disposed to the environmental concepts embedded within the agri-environment measures. This would increase the need to police the operation of these measures and lead to Member States incurring higher costs in operating them.[114]

A number of options, therefore, existed for the development of cross-compliance measures and for linking these measures to the direct payments that had now become an integral part of agricultural production policy. However, it should also be noted that cross-compliance measures would be heavily dependent upon the continued use of direct payments within the European Community's agricultural policies. The continuation of these payments will also, to a large extent, depend

---

110   See Baldock, D., Mitchell, K., footnote 98 above, 69.

111   See Taylor J.P., Dixon, J.B. (1990), *Agriculture and the Environment: Towards Integration*, (Sandy: RSPB) 21.

112   These agri-environment policies will be examined in Chapter 5.

113   See Baldock, D., Mitchell, K., footnote 98 above, 71.

114   *Ibid.* at 69.

upon future agreements within World Trade talks.[115] If direct payments to farmers were, in future, abolished then cross-compliance would at best have represented a short-term environmental remedy. Ironically, the use of cross-compliance measures might also be used to make the continuation of direct payments more defensible within future World Trade talks. Their existence could enhance the environmental qualities of those payments.

Developing Cross-Compliance Methods within the Common Agricultural Policy

In 1999, the European Commission observed that:[116]

> The philosophy underpinning the environmental aspects of CAP reform is that farmers should be expected to observe basic environmental standards without compensation. However, wherever society desires that farmers deliver an environmental service beyond this base line level, this service should be specifically purchased through agri-environmental measures.

The reference to a requirement that farmers should be expected to observe basic environmental standards without compensation would suggest that the European Community favoured a red ticket approach to cross-compliance. In reality, the Member States themselves proved reluctant to endorse this approach. For example, in 1992 the United Kingdom sought to persuade other Member States that all direct payments to farmers should be linked to compulsory obligations.[117] In practice, all that could be agreed was that Member States should have discretion to link such conditions to the direct payments made to beef and sheep farmers.[118] In the United Kingdom this resulted in the introduction of measures to tackle environmental damage resulting from overgrazing and the use of unsuitable techniques to feed animals.[119] The Ministry of Agriculture, Fisheries and Food ('MAFF'), as it was then, could withhold direct payments from farmers who failed to observe restrictions that had been placed upon their livestock numbers or who continued to use unsuitable supplementary feeding methods that were causing environmental

---

115   See, for example, Potter, C. (1996), *Decoupling by Degrees? Agricultural Liberalisation and its Implications for Nature Conservation in Britain*, Research Paper 196, (Peterborough: English Nature) 13.

116   European Commission (1999), *Directions Towards Sustainable Agriculture*, COM (1999) 22 Final 20.

117   Baldock, D., Beaufoy, G. (1992), *Plough On! An Environmental Appraisal of the Reformed CAP* (Godalming: World Wide Fund for Nature) 64.

118   See, respectively, Council Regulation 3611/93 ([1993] OJ L328/7) and Council Regulation 233/94 ([1994] OJ L30/9).

119   See the Beef Special Premium Regulations 1993, SI1993/1734, the Suckler Cow Premium Regulations 1993, SI 1993/1441, the Sheep Annual Premium Regulations 1992, SI 1992/2677 and the Hill Livestock (Compensatory Allowances) Regulations 1994, SI 1994/2740.

damage.[120] Elsewhere in the European Community, only Greece also chose to act upon the discretion to introduce cross-compliance measures.[121]

In 1999, the Agenda 2000 reforms introduced the requirement that 'appropriate' environmental conditions should be attached to all direct payments made to farmers. However, once again, much was left to the Member States. Regulation 1259/99 provided that where direct payments were made to farmers:[122]

> ...Member States shall take the environmental measures they consider to be appropriate in view of the situation of the agricultural land used or the production concerned and which reflect the potential environmental effects. These measures may include:
> - support, in return for agri-environmental commitments,
> - general mandatory environmental requirements,
> - specific environmental requirements constituting a condition for direct payments.

The Regulation also went on to provide that Member States should decide upon and apply penalties that were appropriate and proportionate to the seriousness of the ecological consequences of not complying with the measures that were adopted.[123]

In proposing these measures, the European Commission observed that 'cross-compliance has a great potential, if implemented by the Member States, to contribute to environmental improvement and sustainable development in agriculture.'[124] Unfortunately, the 1999 reforms failed to put in place effective cross-compliance measures. This was principally due to the level of discretion provided to Member States. Council Regulation 1259/99 provided three methods by which environmental protection requirements could be integrated into agricultural production policy. Two of these were cross-compliance based measures. On the one hand, Member States could make the availability of direct payments conditional upon the fulfilment of specific environmental conditions. On the other hand, Member States could require recipients to comply with the general environmental laws of their Member State. However, the third option (support, in return for agri-environmental commitments) enabled Member States to decide that appropriate

---

120   Each Regulation defines 'unsuitable supplementary feeding methods' as meaning 'providing supplementary feed (other than to maintain livestock during abnormal weather conditions) in such a manner as to result in damage to vegetation through excessive trampling or poaching of the land by animals or excessive rutting by vehicles'.

121   See Court of Auditors, *Greening the CAP*, [2000] OJ C353/1 at 11.

122   Council Regulation 1259/99 [1999] OJ L160/113 Article 3. See also Cardwell, M., The Polluter Pays Principle in European Community Law and its Impact on UK Farmers, [2006] 59 *Oklahoma Law Review* 89.

123   *Ibid.*

124   European Commission, *Directions Towards Sustainable Agriculture*, COM (1999) 22 final 21.

environmental protection would be achieved simply through farmer participation in the voluntary environmental land management contracts that formed part of their rural development programmes. This latter option was similar to the green ticket option referred to above. Farmers were able to supplement their income by participating in these voluntary schemes in addition to receiving direct payments.

The European Commission subsequently reported that, in environmental terms, the measures taken by Member States to fulfil the requirements of Regulation 1259/99 left 'considerable scope for improvement'.[125] In particular, they observed that little use had been made of the cross-compliance mechanisms.[126] Other observers have similarly noted that only limited use was made of the option to link the availability of direct payments to particular environmental conditions.[127] In practice, Member States each had very different priorities in relation to combating environmental pollution and had already adopted a variety of measures to deal with this issue.[128] Equally, other Member States failed to adopt cross-compliance measures because they feared that they might lead to them adopting tougher environmental measures than those applied in other Member States and, consequently, reducing the competitiveness of their national agricultural industries.[129] Indeed, the strong political influence exerted by farming organisations in some Member States, also contributed to this situation. Collectively, these issues could only be addressed by a measure that introduced compulsory cross-compliance and which made similar demands of farmers in all Member States. This, ultimately, was what the European Commission sought to achieve in the 2003 Mid Term Review of Agriculture.

---

125   European Commission, *Mid Term Review of the Common Agricultural Policy*, COM (2002) 394 final 8.

126   *Ibid.*

127   See Background Paper for 'Policy Forum on Cross Compliance in the CAP' prepared by the Institute for European Environmental Policy in co-operation with the German Federal Research Centre, Dutch Centre for Agriculture and the Environment, Spanish Universidad Politecnica de Madrid, Czech Institute for Structural Policy, Danish Royal Veterinary and Agricultural University, the Agricultural University of Athens, Lithuanian Institute of Agrarian Economics and Italian Instituto Nazionale di Economia Agraria. The back ground paper can be accessed at http://www.ieep.org.uk [last accessed 6 August 2009].

128   See, for example, Baldock, D., Bennett, G. (1991), *Agriculture and the Polluter Pays Principle* (London: Institute for European Environmental Policy), which compares pollution control measures adopted in Belgium, Denmark, France, Germany, the Netherlands and the United Kingdom.

See also Hawke, N. Implementation and Enforcement of Agri-Environmental Measures, in Van Dunné, J.M. (ed.) (1996), *Non Point Source River Pollution: The Case of the River Meuse* (London: Kluwer Law International) 75, which compares the implementation of environmental measures in the United Kingdom and Denmark in relation to water pollution by nitrates, set aside and soil quality.

129   Baldock, D., Bennett, G. footnote 128 above at 24.

The European Commission's proposals for the 2003 Mid Term Review of Agriculture finally sought to provide a more central role for cross-compliance within agricultural production policy.[130] They proposed that all farmers who received direct payments should be required to meet mandatory environmental standards in relation not just to the environment, but also food safety, animal welfare and occupational safety.[131] The Commission proposals, as amended, were ultimately adopted by the Council as Regulation 1782/2003.[132] Under this Regulation, farmers were required to comply with a range of European Community legislation in order to remain eligible to receive direct payments. From 1 January 2005, farmers had to ensure that their farming practices complied with a number of Community measures on the environment, public and animal health and the identification and registration of animals. On 1 January 2006, additional Community legislation on public, animal and plant health was also added. Additionally, since 1 January 2007 farmers have also been required to comply with a number of Community directives on animal welfare. Collectively, these measures promote greater uniformity of practice, since farmers throughout the Community are required to ensure that their practice complies with the same Community legislation.

Regulation 1782/2003 introduced a further condition. In order to remain eligible to receive direct payments, farmers were also required to maintain their farms in 'good agricultural and environmental condition'. This obligation took effect from 1 January 2005 and required farmers to ensure that they, at least, observed minimum conditions established by their national or regional agricultural authority. The Regulation provided a framework of issues, which national administrations were required to address in their codes of agricultural good practice.[133]

More recently, the 2008 Health Check has retained the structure of the compulsory cross-compliance requirements introduced by Council Regulation 1872/2003. It has, however, streamlined the list of Community legislation that farmers must comply with.[134]

Cross-Compliance in Present Production Policy

Following the 2008 Health Check, farmers across the European Community must comply with the following EC legislation in order to remain fully eligible for direct payments under the Common Agricultural Policy:

---

130   European Commission, footnote 125 above, 8. See also Cardwell, M. footnote 122 above, 105.

131   *Ibid.* See also European Commission, *A Long Term Policy Perspective for Sustainable Agriculture*. COM (2003) 23 Final.

132   Council Regulation 1782/2003 establishing common rules for direct support schemes under the common agricultural policy and establishing certain support schemes for farmers [2003] OJ L270/1.

133   Council Regulation 1782/2003, Article 5 and Annex IV.

134   See Council Regulation 73/2009, [2009] OJ L30/16, Article 4 and Annex II.

Environmental Protection
- Council Directive 79/409/EEC on the conservation of wild birds, [1979] OJ L103/1: [Articles 3(1), 3(2)(b), 4(1), 4(2), 4(4) and 5(a), (b) and (c)]
- Council Directive 80/68/EEC on the protection of groundwater against pollution caused by certain dangerous substances, [1980] OJ L20/43, [Articles 4 and 5]
- Council Directive 86/278/EEC on the protection of the environment, and in particular of the soil, when sewage sludge is used in agriculture, [1986] OJ L181/6, [Article 3]
- Council Directive 91/676/EEC on the protection of waters against pollution caused by nitrates from agricultural sources, [1991] OJ L375/1, [Articles 4 and 5]
- Council Directive 92/43 on the conservation of natural habitats and of wild flora and fauna, [1992] OJ L206/7, [Articles 6 and 13(1)(a)]

Public and Animal Health and the Identification and Registration of Animals
- Council Directive 2008/71/EC on identification and registration of pigs, [2008] OJ L213/31, [Articles 3–5]
- Regulation 1760/2000 of the European Parliament and of the Council establishing a system for the identification and registration of bovine animals and regarding the labelling of beef and beef products, [2000] OJ L204/1, [Articles 4 and 7]
- Council Regulation 21/2004 establishing a system for the identification and registration of ovine and caprine animals, [2004] OJ L5/8, [Articles 3–5]

Public, Animal and Plant Health
- Council Directive 91/414/EEC on the placing of plant protection products on the market, [1991] OJ L230/1, [Article 3]
- Council Directive 96/22/EEC on the prohibition of the use in stockfarming of certain substances having a hormonal or thyrostatic action and of beta-agonists, [1996] OJ L125/3, [Articles 3(a), (b), (d) and (e) and 4, 5 and 7]
- Regulation 178/2002 of the European Parliament and of the Council laying down the general principles and requirements of food law, establishing the European Food Safety Agency and laying down procedures in matters of food safety, [2002] OJ L31/1, [Articles 14, 15, 17(1) and 18–20]
- Regulation 999/2001 of the European Parliament and of the Council laying down rules for the prevention, control and eradication of certain transmissible spongiform encephalopathies, [2001] OJ L147/1, [Articles 7,11–13 and 15]

Notification of Diseases
- Council Directive 85/511/EEC introducing Community measures on the control of foot and mouth disease, [1985] OJ L315/11, [Article 3]

- Council Directive 92/119/EEC introducing general Community measures for the control of certain animal diseases and specific measures relating to swine vesicular disease, [1992] OJ L62/69, [Article 3]
- Council Directive 2000/75/EC laying down specific provisions for the control and eradication of bluetongue, [2000] OJ L327/74, [Article 3]

Animal Welfare
- Council Directive 91/629 laying down minimum standards for the protection of calves, [1991] OJ L340/28, [Articles 3–4]
- Council Directive 91/630 laying down minimum standards for the protection of pigs, [1991] OJ L340/33, [Articles 3 and 4(1)]
- Council Directive 98/58 concerning the protection of animals kept for farming purposes, [1998] OJ L221/23, [Article 4]

In addition, Regulation 1872/2003 also required Member States to develop standards of good agricultural and environmental condition that addressed particular issues and standards set out in the Regulation. Farmers will also be required to comply with these standards of practice as part of their cross-compliance obligations. The standards and issues were amended by the 2008 Health Check. In particular, it introduced a requirement that national standards of good agricultural and environmental condition should address the issue of water management and expanded upon land management requirements. The European Commission viewed the introduction of requirements on the issue of water management as a means of retaining the environmental benefits of set aside, following the abolition of this production control tool.[135] Environmental groups, however, have been highly critical of this approach.[136]

The current Community framework for standards of good agricultural and environmental condition is set out in Table 3.1.[137]

In practice, since agriculture is a devolved matter, each United Kingdom region has introduced its own standards of agricultural and environmental condition.[138]

---

135   European Commission, footnote 67 above, 5. The set aside scheme is examined in detail later in this chapter.

136   See later, at footnote 193.

137   As established by Council Regulation 73/2009, [2009] OJ L30/16, Article 6 and Annex III.

138   See, respectively, the Common Agricultural Policy Single Payment and Support Schemes (Cross Compliance) (England) Regulations 2004, SI 2004/3196 as amended by the Common Agricultural Policy Single Payment and Support Schemes (Cross Compliance) (England) (Amendment) Regulations 2005, SI 2005/918; the Common Agricultural Policy Schemes (Cross Compliance) (Scotland) Regulations 2004, SSI 2004/518, the Common Agricultural Policy Single Payment and Support Schemes (Cross Compliance) (Wales) Regulations 2004, SI 2004/3280 (W.284) and the Common Agricultural Policy Single Payment and Support Schemes (Cross Compliance) Regulations (Northern Ireland) 2005, SR 2005/6.

Table 3.1   The Requirements of 'Good Agricultural and Environmental Condition'

Issue	Compulsory Standards	Optional Standards
Soil erosion Protect soil through appropriate measures	Minimum soil cover Minimum land management reflecting site specific conditions	Retain Terraces
Soil organic matter Maintain soil organic matter levels through appropriate practices	Arable stubble management	Standards for crop rotations
Soil Structure Maintain soil structure through appropriate measures		Appropriate machinery use
Minimum level of maintenance Ensure a minimum level of maintenance and avoid the deterioration of habitats	Retention of landscape features, including, where appropriate, hedges, ponds, ditches, trees in line or in group or isolated and field margins. Avoiding the encroachment of unwanted vegetation on agricultural land. Protection of permanent pasture.	Minimum livestock stocking rates or/and appropriate regimes. Establishment and/or retention of habitats. Prohibition of the grubbing up of olive trees. Maintenance of olive groves and vines in good vegetative condition.
Protection and Management of Water	Establishment of buffer strips along water courses. Where use of water for irrigation is subject to authorisation, compliance with authorisation procedures.	

The legality of the United Kingdom approach is the subject of an ongoing judicial challenge. In *R (Horvath) v Secretary of State for the Environment, Food and Rural Affairs*[139], the applicant has objected to the fact that only the standards of good agricultural and environmental condition for England require farmers to observe conditions relating to the upkeep of pathways and bridleways. In contrast, farmers in Wales, Scotland and Northern Ireland do not risk losing some or all

---

139   [2006] EWHC 1833 (Admin.).

of their single payments by failing to maintain footpaths and bridleways. The applicant has argued that Regulation 1872/2003 gave national authorities no power to develop cross-compliance requirements concerning footpaths and bridleways. Additionally, he argued that differences between standards implemented in England and those applying in the other parts of the United Kingdom amount to unlawful and unjustifiable discrimination, contravening the general principles of European Community law.[140] At the time of writing, the proceedings were stayed, pending the outcome of a preliminary reference to the European Court of Justice.[141]

The European Community requires Member States to enforce these cross-compliance provisions by carrying out on-the-spot inspections of at least 1 per cent of all farmers who submit applications for single farm payments.[142] In practice, the inspections are usually not 'on-the-spot'. As long as the statutory measures, upon which cross-compliance is based, do not forbid it and the purposes of the inspection will not be jeopardised, farmers can be given up to 14 days notice of the inspection.[143] However, one exception is that farmers cannot receive more than 48 hours notice of inspections that concern livestock aid.[144] Between 20 and 25 per cent of the farmers inspected should be selected randomly, with the remaining farms selected on the basis of a risk assessment.[145] Member States can analyse risk at the level of individual farms, categories of farms or on the basis of geographical zones.[146] They can also take account of factors such as farmers' participation in farm advisory schemes or in farm certification schemes – such as those that larger food supermarkets typically require their farm suppliers to undergo – where that scheme is relevant to the requirements and standards at issue.[147]

Where an inspection reveals that a farmer has failed to comply with the cross-compliance measures in place, the penalty imposed will depend upon whether this non-compliance is the result of negligence or intent. Equally, it will also be determined by whether the farmer has committed previous breaches. Where it is

---

140   On the general principle of non discrimination see generally, Craig, P., De Búrca, G. (2007), *EU Law: Text, Cases and Materials*, 4th ed. (Oxford: Oxford University Press) 558.

141   The Court of Appeal subsequently rejected an appeal against this reference – see *R (Horvath) v Secretary of State for the Environment, Food and Rural Affairs*, [2007] EWCA Civil 620 or Times Law Reports 30 July 2007.

142   See Commission Regulation 796/2004 laying down detailed rules for the implementation of cross-compliance, modulation and integrated administration and control systems provided for by Council Regulation 1782/2003, [2004] OJ L141/18, as last amended, by Commission Regulation 1124/2008, [2008] OJ L303/7, Article 44. However, Member States are also required to undertake further eligibility inspections in relation to the integrated administration and control system.

143   *Ibid.* Article 23a.

144   *Ibid.*

145   *Ibid.* Article 45(2).

146   *Ibid.*

147   *Ibid.*

a first non-compliance and the result of negligence then, as a general rule, a 3 per cent deduction should be made to the total direct payments made to the farmer in that calendar year.[148] However, where a first non-compliance was intentional, as a general rule 20 per cent of those monies should be deducted.[149] Where the intentional non-compliance relates to a particular aid scheme, the farmer will also be excluded from that aid scheme in the following calendar year.[150]

The penalties set out in the previous paragraph are those which apply as 'a general rule'. In each case, national agricultural authorities are given some discretion in determining the actual penalty that should apply.[151] This discretion must be exercised in the light of the recommendations made in the report of the body which carried out the farm inspection.[152] Their report should indicate the importance of the breach and explain any factors that should be considered in determining whether a higher or lower financial penalty should be applied.[153]

Based upon this report, where the breach was due to negligence, the penalty imposed may be reduced to a 1 per cent deduction or raised to a 5 per cent deduction.[154] Equally, they can decide not to impose any deduction at all.[155] Where an inspection reveals an intentional non-compliance, the national authorities may decide, again on the basis of the inspection report, to reduce the deduction to 15 per cent or to increase it to up to 100 per cent of all direct payments to be made in a calendar year.[156]

In cases of repeat infringements, national authorities may impose greater financial penalties. Where these breaches derive from negligence, these penalties may be up to 15 per cent of the direct payments to be made in a calendar year.[157] Once this ceiling is reached, future infringements will be treated as being intentional and even greater financial penalties imposed.[158] Where repeat infringements are the result of intentional non-compliance, the penalties imposed can, ultimately, rise to 100 per cent of the direct payments that would be due to the farmer in that particular calendar year.[159]

---

148 *Ibid.* Article 66(1).

149 *Ibid.* Article 67(1).

150 *Ibid.* Article 67(2).

151 *Ibid.* Articles 66(1) and 67(1).

152 *Ibid.*

153 *Ibid.* Article 48(1) (c).

154 *Ibid.* Article 66(1).

155 *Ibid.*

156 *Ibid.* Article 67(1).

157 *Ibid.* Article 66(4). Repeated non compliances will be penalised by multiplying the initial penalty by 3, subject to a maximum deduction of 15 per cent.

158 *Ibid.* Article 66(4). Where a further non compliance due to negligence is determined then the penalty will be calculated by multiplying the previous percentage penalty, before the limitation to 15 per cent, by 3.

159 *Ibid.* Article 67(1).

These cross-compliance provisions are certainly capable of ensuring that the European Commission's objective is achieved – that farmers should observe basic environmental standards without compensation. However, this will only be the case if Member States take adequate steps to implement and enforce both the cross-compliance measures and also the European Community legislation upon which they are based. Sadly, one recent evaluation for the European Commission has suggested that Member State engagement has been extremely variable.[160] From an environmental perspective, it should be a matter of concern that the report found that only three Member States (Germany, Italy and the United Kingdom) had actually established obligations for farmers under all articles of the European Community legislation required for cross-compliance.[161] Similarly, Member State practice also varied widely in developing standards of good agricultural and environmental condition. Here the issue that received least attention was soil structure. Some 14 Member States had failed to include any obligations in this area within the standards that they had developed.[162] Equally, the Court of Auditors also recently published a highly critical report on cross-compliance measures.[163] They also noted that Member States had not translated all the cross-compliance standards into obligations that applied at farm level and were critical of weak monitoring of the standards by national agriculture authorities. Such limitations, obviously, limit the environmental potential of the cross-compliance scheme. They also serve to, once again, question Member State commitment to the scheme. Sadly, as in other areas of European Community environmental law, the scheme may only begin to achieve its full environmental potential when the European Commission begins to exercise its enforcement powers against recalcitrant Member States.

The Integration of Environmental Obligations Through Set Aside Measures

An Introduction to Set Aside within the Common Agricultural Policy

The European Community has introduced a number of set aside measures over the last 20 years, providing for farmers to set arable land aside from agricultural

---

160   See Swales, V., Arblaster, K., Bartley, J., Farmer, M. (2007), *Evaluation of Cross Compliance as Foreseen under Regulation 1782/2003*, Alliance Environnement at http://www.ieep.eu/publications/publications.php [last accessed 10 August 2009]. Alliance Environnement comprises the Institute for European Environmental Policy (United Kingdom) and Oréade-Bréche (France).

161   *Ibid.* 21.

162   *Ibid.* 47 and 63. The Member States concerned were: Belgium (Flanders), the Czech Republic, Germany, Denmark, Estonia, Hungary, Lithuania, Luxembourg, Latvia, the Netherlands, Poland, Portugal, Sweden, Slovenia.

163   Court of Auditors (2008) Special Report 8/2008, *'Is Cross Compliance an Effective Policy?*

production. This was done in two different ways. On the one hand, voluntary schemes provided payments to arable farmers who agreed to take a portion of their land out of production. On the other hand, the introduction of direct payments, subsequently, enabled the Community to introduce a compulsory scheme. Arable Farmers were then required to set a portion of their land aside from production in order to remain eligible to receive direct payments.

The European Community introduced set aside measures in 1988, when Regulation 1094/88 introduced a voluntary set aside scheme.[164] Farmers could receive payments in return for taking at least 20 per cent of their arable land out of arable production for at least five years. They also had the right to terminate their agreement after three years. The land withdrawn from production could be left fallow, wooded or used for non-agricultural purposes.[165] The principal objective of this Regulation was reducing surplus production. For example, its preamble noted that 'arrangements for set aside of arable land may assist the various sectors, and in particular those with surplus production, to adjust to market requirements.'[166] In contrast, the scheme's environmental credentials were limited. Member States were required to ensure that any land set aside was maintained in good agricultural condition. The Regulation stipulated that the measures taken 'may include an obligation on the farmer to ensure that the agricultural land withdrawn from production is maintained with a view to protecting the environment and natural resources.'[167] Member States, therefore, had discretion to include environmental protection measures within the conditions imposed on farmers taking part in the scheme. In real terms, the obligation to keep land in good agricultural condition was principally intended as a means to ensure that the land could, in future, be returned to arable production.

In 1991, the Community introduced a second voluntary set aside scheme. On this occasion, farmers entered a temporary set aside scheme for one year – the arable production year 1991–1992.[168] Farmers who took part in the scheme received payments in return for withdrawing at least 15 per cent of their arable land from production.[169] Member States had to 'apply appropriate measures in favour of the environment that correspond to the specific situation of the area set aside'.[170] They, therefore, had an obligation to ensure that participating farmers were subject to environmental conditions, but retained broad discretion in developing these conditions.

---

164   Council Regulation 1094/94 [1988] OJ L106/28.

165   *Ibid.* Article 1.

166   *Ibid.*

167   *Ibid.*

168   Council Regulation 1703/91 introducing a temporary set aside scheme for arable land for the 1991/92 marketing year, Article 1(3).

169   *Ibid.*

170   *Ibid.* Article 1(4) (b).

The reform of the common organisation in cereal production, in 1992, under Regulation 1765/92,[171] saw the European Community balance the introduction of arable area payments with the introduction of a compulsory set aside regime. Council Regulation 1765/92 required Member States to submit plans to the European Commission in which they identified the separate arable regions that would be recognised within their territories.[172] Average yields per hectare were then calculated for each region. This calculation was made on the basis of the yields that had been produced in the marketing years 1986/87 to 1990/91, with the years with the highest and lowest yields being excluded.[173] Based upon this average regional yield per hectare, farmers who claimed arable area payments upon an area greater than that required to produce 92 tonnes of cereals were required to set a fixed percentage of their arable land aside from arable production.[174] Alternatively, they could use this set aside land to grow crops that would be used within the Community to manufacture products not intended for human or animal consumption. The Regulation also enabled Member States to exempt small farmers, those claiming arable area payments on an area no bigger than that required to produce 92 tonnes of cereals, from these set aside obligations.[175]

In setting the percentage of land to be compulsorily set aside from arable production, the Community initially distinguished between rotational set aside, where the arable land set aside from production would be rotated around the farm and each parcel of land could only be set aside once in each five year cycle, and non-rotational set aside, where the same parcel of land would be set aside from production for five successive marketing years.[176] At first, farmers were required to set aside at least 15 per cent of their arable land if they opted for rotational set aside and at least 18 per cent where non-rotational set aside was chosen. This distinction remained until 1996 when a single set aside rate was adopted for all set aside land.

Regulation 1765/92 required Member States to 'apply appropriate environmental measures which correspond to the specific situation of the set-aside land.'[177] Subsequently, this regulation was repealed by Regulation 1251/99, which continued to operate the compulsory set aside scheme and made identical provision for the environmental management of set aside land.[178] Once again therefore, whilst Community law made it compulsory for Member States to regulate the

---

171   Council Regulation 1765/92 [1992] OJ L270/1.
172   *Ibid.* Article 3.
173   *Ibid.* Article 3(2).
174   *Ibid.* Articles 7 and 8.
175   *Ibid.* Article 8.3.
176   *Ibid.* Article 7 provided for both rotational and non-rotational set aside to operate. The time periods that applied to each method of set aside were initially established by Commission Regulation 762/94.
177   Council Regulation 1765/92, Article 7(3).
178   Council Regulation 1251/99, [1999] L160/1, Article 6(2).

environmental management of set aside land it also provided very broad discretion as to the measures that they put in place. In essence, however, this made arable set aside the European Community's first compulsory cross-compliance measure.

Compulsory set aside remained a part of production policy until 2007. That year, the Council of the European Union voted to reduce the obligatory set aside rate to zero. More recently, the 2008 Health Check reform legislation abolished compulsory set aside all together. The abolition of compulsory set aside resulted from a number of factors – such as a growing demand for cereal crops on international markets, a corresponding reduction in surplus stocks held by European Community intervention agencies and the increasing demand for non-food crops to be grown as a biofuel source.[179] In relation to biofuels, for example, the European Community's Biofuels Directive initially provided that biofuels should make up 2 per cent of all petrol and diesel consumption within the Community by 2005 and 5.75 per cent by 2010.[180] More recently, in 2007, the European Council endorsed European Commission proposals that biofuels should account for at least 10 per cent of all petrol and diesel used in transport within the Community by 2010.[181] The Community's decision to abolish compulsory set aside requirements, therefore, reflected the changing market place, which would enable set aside land to be used to satisfy growing demand for food and energy crops.

The European Community's compulsory set aside scheme was primarily motivated by a desire to control production levels and environmental management of the set aside land was very much of secondary concern. This, however, was not the case with one further set aside measure that the Community also introduced in 1992. Regulation 2078/92 on agricultural production methods compatible with the requirements of the protection of the environment and maintenance of the countryside made provision for Member States to introduce measures to enable all farmers, not just arable farmers, to engage in the long term set aside of agricultural land for environmental reasons.[182] Under this scheme farmers could receive payments in return for agreeing to set farmland aside from production for at least 20 years and to use that land for environmental purposes, 'in particular for the establishment of biotope reserves or national parks or for the protection of hydrological systems.'[183] The Regulation also made provision for farmers to be paid to manage this land for public access and leisure use.[184]

---

179   See generally, The Silent Tsunami: The food crisis and how to solve it, *The Economist* Volume 387, number 8576.

180   Council Directive 2003/30/EC, [2003] OJ L123.

181   See variously European Commission, *An EU Strategy for Biofuels*, COM(2006) 34 final, European Commission, *Renewable Energy Road Map – Renewable Energies in the 21st Century: Building a More Sustainable Future*, COM(2006) 848 final.

182   Council Regulation 2078/92, [1992] OJ L215/85.

183   Article 2(1) (f).

184   *Ibid.* Article 2(1) (g).

The Environmental Credentials of the Common Agricultural Policy's Set Aside Measures

The Community's initial voluntary set aside schemes, those introduced in 1988 and 1991, provided very few environmental benefits. This is, perhaps, unsurprising since they were almost entirely motivated by economic considerations. In reality, the initial measures also had limited success in reducing arable production. In the United Kingdom, for example, it has been estimated that both schemes, together, resulted in just under 5 per cent of arable land being taken out of production.[185] Even where land was set aside, environmentalists criticised these schemes for allowing fallow fields to degenerate without proper management and for encouraging a proliferation of non-agricultural development, such as golf courses.[186]

Ultimately, however, the initial voluntary schemes proved to be testing grounds for the much broader compulsory measures introduced in 1992. The introduction of compulsory set aside had the potential to dramatically increase the amount of arable land set aside from production. In practice, the amount of land actually set aside varied according to both the set aside rate established by the Council of the European Union and also the farm structure of individual Member States. In terms of the set aside rate, this had initially been set at 15 per cent, for rotational set aside, but this rate later varied between 12 per cent and 5 per cent before being set at zero in 2007.[187] These variations, obviously, influenced the amount of land set aside. Equally, variations in farm structure were also important. The compulsory set aside scheme enabled Member States to exempt small producers, those claiming arable area payments on an area greater than that required to produce 92 tonnes of cereals. In England, average the farm size is approximately twice the European Community average. As a result, most arable farmers in England faced compulsory set aside obligations, resulting in a large area of arable land being set aside. For example, set aside accounted for 500,000 hectares of agricultural land in England in 1995–1996 – 11 per cent of all eligible arable land.[188] In contrast, the scheme had little impact in other Member States where arable farms were smaller.

---

185   MAFF press release 456/91 quoted in Rodgers, C. Environmental Gain, Set Aside and the Implementation of EU Agricultural Reform in the United Kingdom, in Rodgers (ed.) (1996), *Nature Conservation and Countryside Law* (Cardiff: University of Wales Press) 116.

186   Council for the Protection of Rural England (1993), *Agriculture and England's Environment* (London: CPRE) paras. 6.1–6.4, quoted in Rodgers, C., footnote 185 above.

187   In relation to the early variations see Hawke, N., Kovaleva, N. (1998), *Agri-Environmental Law and Policy* (London: Cavendish Publishing Ltd) 85.

188   See Crabb, J. et al., Set Aside Landscapes: Farmer Perceptions and Practices in England, [1998] 23 *Landscape Research* 237. See also Winter, M., The Arable Crops Regime and the Countryside Implications, in Brouwer, F., Lowe, P. (eds) (2000), *CAP Regimes and the European Countryside* (Wallingford: CABI Publishing) 121.

For example, it accounted for 1 per cent of arable land in Greece, 2 per cent in the Netherlands and 4 per cent in Belgium.[189]

The European Community took greater care to introduce environmental requirements into the compulsory set aside scheme, in comparison to the previous voluntary models. In practice, at farm level, environmental outputs depended upon whether a farmer chose rotational or non-rotational set aside. As noted above, rotational set aside resulted in different parcels of land being set aside each year, whereas in non-rotational set aside the same land was set aside over a five-year cycle. In practice, under the European Commission's implementing regulation, rotational set aside only required land to be set aside for a minimum of seven months.[190] Consequently, the longer term management of land subject to non-rotational set aside was likely to produce greater environmental benefits. However, set aside was, in reality, a production-control tool, designed to enable the Community to adjust production levels to restrict surplus production. As rotational set aside enabled producers to respond more flexibly to market demands, it was the preferred model of both the European Community and farmers.[191]

Nevertheless, despite the criticisms above, set aside did bring environmental benefits. This has been the case, particularly, in relation to farmland bird biodiversity. Several studies have shown that rotational set aside provided important food sources or nesting grounds for a range of farmland bird species in the United Kingdom.[192] Consequently, conservation groups such as the RSPB and Birdlife International have warned of the danger of significant decline amongst these species if adequate measures are not taken to retain the environmental benefits achieved by compulsory rotational set aside.[193] This warning has also

---

189   J. Crabb et al., footnote 188 above, quoting figures published in Hansard volume 295, col. 103 on 3rd June 1997. See also Winter, M., footnote 188 above, 121.

190   European Commission Regulation 762/94, [1994] OJ L90, Article 3(4).

191   For example, the preamble to Council Regulation 1765/92, which introduced compulsory set aside, states that 'the set aside should normally be organised on the basis of a rotation of areas: whereas non-rotational fallow should be permitted.'

192   See Council for the Protection of Rural England and World Wide Fund for Nature (1996), *Sustainable Agriculture in the UK* (London: CPRE & WWF) 116, Crabb, J., footnote 200 above, Firbank, L.G. (ed.) (1998), *Agronomic and Environmental Evaluation of Set Aside under the EC Arable Area Payments Scheme*, Volume 1 (Grange-over-Sands: Institute of Terrestrial Ecology). Poulsen J.G., Sotherton N.W., Aebischer N.J. (1998), Comparative nesting and feeding ecology of skylarks Alauda arvensis on arable farmland in southern England with special reference to set-aside, [1998] *Journal of Applied Ecology* 131.

193   See RSPB Press Release 'Loss of Set Aside Threatens Farmland Bird Recovery', at http://www.rspb.org.uk/news/details.asp?id=tcm:9-182116 [last accessed 26 June 2008] and Birdlife International Press Release 'Don't set aside set-aside: Europe's nature under threat as Commission decides to reduce set-aside to 0%' at http://www.birdlife.org/news/pr/2007/09/set-aside.html [last accessed 26 June 2008].

been repeated by England's nature conservation regulator, Natural England.[194] The European Commission, in its proposals for the 2008 Health Check, suggested that reforms to Community cross-compliance and rural development policy measures would ensure that the environmental benefits of set aside were retained.[195] The reference to cross-compliance concerns the fact that the Commission's proposals extended the requirements of good agricultural and environmental condition that Member States should apply, to require farmers to establish buffer strips alongside water courses. Although this was, ultimately, adopted by the Community as part Council Regulation 73/2009, environmental groups have argued that, whilst buffer strips alongside water courses may prevent nitrate pollution of water, they will do little to support biodiversity.[196] Equally, question marks hang over the use of rural development plans. This is a reference to the use of environmental land management contracts to protect farmland bird habitats. In practice, this will only be successful if Member States provide adequate financial support for these contracts and target them for this purpose. As Chapter 5 illustrates, Member States' past record in using these contracts to deliver environmental goals has been quite mixed. Meanwhile, in England, DEFRA has worked with a number of industry groups to introduce voluntary measures designed to encourage farmers to retain environmental benefits achieved through set aside.[197]

Perhaps of all the set aside schemes, it was the 1992 introduction of 20-year environmental set aside that, at face value, offered the greatest potential for long environmental benefits. In the United Kingdom, this measure was implemented through the introduction of a Habitat Improvement Scheme which promoted the creation of wildlife habitats or the improvement of existing habitats.[198] In England, the scheme targeted water fringe habitats along six designated water bodies[199] – recognised salt marshes in south east England[200] and land that had been set aside under the Community's initial five year voluntary set aside programme.[201] In reality, the scheme was not a success. It operated on a voluntary basis and, by

---

194   See Natural England press release, 'We must not set our natural environment aside', at http://www.naturalengland.org.uk/about_us/news/2009/050609.aspx [last accessed 10 August 2009].

195   European Commission, footnote 67 above, 9.

196   See ENDS Report 401, June 2008, 52.

197   See 'Benn backs farmers' green offer', http://www.defra.gov.uk/news/2009/ 090709a.htm [assessed August 10th 2009]. The groups involved included DEFRA, the National Farmers Union, the Countryside Land and Business Association, Natural England and the RSPB. The steps to be taken include promoting farmer participation in the Environmental Stewardship agri-environment scheme (see Chapter 5) and encouraging retention of uncultivated buffer strips near water bodies.

198   See Rodgers, C. (1998), *Agricultural Law* 2nd ed. (London: Butterworths) 435.

199   Implemented by the Habitat (Water Fringes) Regulations 1994, SI 1994/1291.

200   Implemented by the Habitat (Saltmarsh) Regulations 1994, SI 1994/1293.

201   Implemented by the Habitat (Former Set Aside) Regulations 1994, SI 1994/1292.

the end of the 1995/96 marketing year, only 301 farmers had enrolled in England, entering some 5,100 hectares of farmland.[202] As Winter has observed, farmers tended to view set aside, in agricultural management terms, as providing land with a break from production, rather than as a means to protect environmentally significant areas.[203] Consequently, farmers were generally reluctant to enrol land for a 20-year programme, a feeling that was compounded by the, unattractively, low payment levels that were on offer under these schemes. Not for the first time, environmental protection was in competition with production policy, where the financial incentives were more attractive.[204]

Agricultural Production Policy Reforms: A Missed Opportunity?

The European Community's agricultural policy has come a long way in recent years. The current policy is quite far removed from the initial one, that has previously been identified as having encouraged so much environmental damage. Today the cross-compliance measures provide a common framework, ensuring that farmers throughout the Community comply with the same legislative obligations. In doing so, the Community has acted to remove much of the scope for unfair competition that existed when Member States simply had discretion to introduce cross-compliance measures. Equally, the codes of agricultural good practice, similarly, provide a common framework for environmental measures, whilst subsidiarity is expected, by enabling regional authorities, to tailor them to local conditions. Thus, the Community can avoid the pitfalls experienced with livestock quotas, where centrally-established measures took no account of the environmental conditions that actually exist in particular areas. However, the current cross-compliance measures do not, by any means, provide a complete solution to the problem of integrating environmental protection measures into agricultural production policy. One major weakness of these cross-compliance measures is that they only operate in areas of agricultural production in which farmers receive direct payments through the Common Agricultural Policy. Although farmers engaged in arable, beef and dairy farming do receive these payments, there are, equally, farmers in other production sectors, such as pig and poultry farmers, who do not. Cross-compliance measures, therefore, do not reach the entire farming community and will have little relevance for farmers who are in this latter position.

Equally, current agricultural production policies could also be viewed as being a lost opportunity. Compulsory cross-compliance measures may now seek to ensure that farmers who receive direct payments do meet certain minimum

---

202   House of Commons Agriculture Select Committee (1997), *Environmentally Sensitive Areas and Other Schemes Under the Agri-Environment Regulation, Volume 1,* Session 1996–1997 Second Report, HC-45, xxi.

203   Winter, M., footnote 188 above, 121.

204   House of Commons Agriculture Committee, footnote 202 above, xxxi.

environmental standards. However, the criticism might be made that these direct payments, themselves, are simply compensating farmers for the impact of agricultural reforms in reducing market guarantees. In other words, that they are compensatory payments that require little in return from their recipients. In 2004, for example, the Community spent €47 billion on the Common Agricultural Policy, some 47 per cent of its total budget.[205] Some 61 per cent of Community expenditure on agriculture was used to fund these direct payments. Seen from an environmental perspective, it might be argued that this money could have been better spent purchasing environmental services from farmers. Indeed, several proposals have been put forward for payment systems which would have operated on this basis.[206] Equally, other reform proposals have also been put forward, by economists, which are not primarily based upon agriculture's role in delivering environmental services.[207]

Perhaps the most influential of the reform proposals, linking direct payments to the provision of environmental services, was the so called 'Buckwell report'.[208] This was a study, conducted for the European Commission's Directorate General on Agriculture, by a group of agricultural economists and rural sociologists drawn from nine European Community Member States. The study advocated the introduction of 'Environmental and Cultural Landscape Payments' to replace the compensatory direct payments that are currently made. It advocated the reduction of commodity prices within the European Community to the levels prevailing upon world markets.[209] Payments, such as the current Single Farm Payment, would be replaced by 'transitional adjustment assistance', a temporary and gradually declining payment designed to help farmers to adjust to the new policy.[210] As this transitional payment faded out, 'Environmental and Cultural Landscape Payments'

---

205   European Commission (2005), *The Agricultural Situation in the European Union: 2004 Report* (Luxembourg: Office for Official Publications of the European Communities), table 3.4.1.

206   See Buckwell, A. et al. (1997), *Towards a Common Agricultural and Rural Policy for Europe* (Luxembourg: Office for Official Publications of the European Communities); Potter C., (1996), *Decoupling by Degrees: Agricultural Liberalisation and its Implications for Nature Conservation in Britain*, Peterborough: English Nature; Jenkins, T.N. (1990), *Future Harvests: The Economics of Farming and the Environment – Proposals for Action* (London: Council for Protection of Rural England).

207   See, for example, Tangermann, S. A Bond Scheme for Supporting Farm Incomes, in Marsh, J. et al. (eds), (1991), *The Changing Role of the Common Agricultural Policy: The Future of Farming in Europe* (London: Belhaven), Harvey, D.R. Agriculture and the Environment: The Way Forward?, in N. Hanley (ed.) (1991), *Farming and the Countryside: An Analysis of External Costs and Benefits* (Wallingford: CAB International).

208   The report is so-called because the research group which wrote the report was chaired by Professor Allan Buckwell of Wye College, University of London.

209   A. Buckwell et al., footnote 218 above, at 14.

210   *Ibid.* at 16.

would play an increasing role.[211] Ultimately, aside from the 'Environmental and Cultural Landscape Payments' only limited market stabilisation measures would also be available, providing intervention purchasing when exceptional circumstances caused uncontrollable market fluctuations.[212]

The study proposed that Environmental and Cultural Landscape payments should be based upon three tiers.[213] All farmers would be required to observe basic environmental standards set out in tier 0, without payment. This would have been similar to the compulsory cross-compliance measures actually put in place in 2003, with farmers expected to meet the standards set by directive such as the Nitrates Directive, the Wild Birds Directive and the Habitats Directive. Additionally, however, farmers who agreed to enter voluntary management agreements would, under tier 1 of the scheme, be entitled to receive Environmental and Cultural Landscape payments in return for maintaining 'high nature value farming systems'. The proposed measures envisaged that the characteristics of high nature value farming systems would be defined at national or regional level in each Member State, based upon a menu provided in a European Community framework regulation. Farmers would have to enter a management agreement, with national agriculture authorities, which would require them to adhere to prescribed environmental management practices.[214] Additionally, the national agriculture authorities would also have been able to designate particular landscapes as being of additional environmental or cultural importance. Again, these areas would be identified on the basis of a menu established in a European Community framework regulation. Examples, alluded to in the study, included 'the protection of particular high interest eco-systems, wetlands, habitats for specific birds or other fauna and the preservation of valued physical features in the rural landscape.'[215] Under tier 2 of the proposed scheme, farmers would have been entitled to additional payments in return for accepting additional management obligations to manage and protect these features. The study also proposed that tier 2 would also be used to provide payments to farmers who agreed to undertake organic farming techniques.[216]

Elsewhere, Jenkins also envisaged a reform under which payments for environmental services, 'Environmental Management Payments', would replace compensatory direct payments.[217] He envisaged these Environmental Management Payments being available, on a voluntary basis, to all farmers throughout the European Community. Farmers would have been required to enrol their whole

---

211   *Ibid.* at p.15.
212   *Ibid.* at p.14.
213   *Ibid.* at 69–75.
214   *Ibid.* at p.70.
215   *Ibid.* at p.74.
216   *Ibid.* at p.74.
217   T.N. Jenkins, footnote 206 above.

farms into the proposed scheme. Essentially, the proposals were based upon the principle that: [218]

> Farmers should have a choice between producing agricultural output for the market or producing environmental output in return for EMPs [Environmental Management Payments] as well as receiving the full market returns from the food and other products produced on the farm.

As with Buckwell, Jenkins' proposed scheme would have centred around a European Community regulation that provided a flexible framework, to take account of the differing agricultural, climatic and environmental conditions throughout the Community.[219] In the United Kingdom, for example, he foresaw that the scheme would differentiate between upland and lowland farms and would, principally, seek to protect wildlife habitats and both landscape and archaeological features.[220] In terms of wildlife habitats, the proposals would have required farmers to retain features such as hedges and ponds along with semi-natural grasslands and moorlands.[221] To protect the landscape, land enrolled within the scheme would have to be maintained in farming and not abandoned or used for building development.[222] As regards archaeological features, there would have been a requirement that archaeological sites should be protected and appropriately managed.[223] These requirements would have provided a basic level of environmental protection under the scheme. Payments to farmers would have consisted of two elements. There would be a fixed payment based upon the hectarage of participating farms.[224] To this would be added additional payments calculated on the basis of factors such as the length of traditional field boundaries that were being maintained on each farm, the number or area of specific wildlife habitats or landscape features being maintained and the number or area of archaeological features that were being maintained.[225] Farmers whose farms were identified as containing high grade conservation land would also have been required to meet additional environmental criteria and would be entitled to receive higher payments.[226] The proposal also envisaged that additional payments would have been made available to farmers who agreed to undertake certain positive environmental works, such as protecting

---

218   *Ibid.* at p.51.
219   *Ibid.* at p.50.
220   *Ibid.* at p.59.
221   *Ibid.* at p.61.
222   *Ibid.* at p.60.
223   *Ibid.* at p.60.
224   *Ibid.* at p.65.
225   *Ibid.* at p.66.
226   *Ibid.* pp. 63, 67.

existing wildlife habitats by creating buffer zones or wildlife corridors or by re-creating lost habitats or landscapes or providing public access to farmland.[227]

Another proposal suggested that the existing compensatory direct payments made to farmers should not be entirely removed.[228] Instead, they should be replaced by a lower basic payment made on an area basis. Farmers could then choose to supplement their incomes by enrolling in voluntary environmental schemes which would be available on a tiered basis. Again, under this proposal, the level of the environmental payments received by farmers would have increased as farmers agreed to observe more stringent environmental conditions set out in higher tiers.

Ultimately, however, the European Community has not chosen to follow any of these suggested approaches. Instead, direct compensatory payments continue to be made to farmers and, indeed, these payments have increased in size under each recent amendment to the Community's agricultural production policies. In this situation, compulsory cross-compliance has been the vehicle chosen to integrate environmental protection measures into agricultural production policy. However, the Community's present measures do actually make provision for farmers to receive payments in return for providing environmental services under 'Environmental Land Management Contracts'. The development of these contracts will be explored in Chapter 5. As such, the Community has put in place a system which, perhaps, goes halfway towards meeting the proposals made in the Buckwell report. Equally, however, the continuation of direct compensatory payments for farmers represents a huge cost to the Community budget. The abolition of these payments would, potentially, have made a much larger sum of Community money available to assist in funding environmental land management contracts.

---

227   *Ibid.* at pp. 58 and 63.

228   See Potter, C. (1996), *Decoupling by Degrees? Agricultural Liberalisation* and *its Implications for Nature Conservation in Great Britain*, English Nature Research Report 196 (Peterborough: English Nature), 6. See also Potter, C. (1998), *Against the Grain, Agri-Environmental Reform in the United States and the European Union* (Wallingford: CAB International) 128–153.

Chapter 4

Environment and Rural Development Policy

Greening Rural Development Policy

Chapter 1 highlighted that the European Community's current rural development policy, originally developed as a mechanism to promote structural change in agriculture. Policy makers subsequently realised that agriculture was no longer the predominant employer in most rural areas. As a result, the European Community transformed its agricultural structural policy into a more broadly-based rural development policy. This change was initially introduced by Regulation 1257/99,[1] with the aim of supporting wider economic and social development in rural areas, as a whole. The transformation from agricultural structural policy to rural development policy was, subsequently, given further emphasis by Regulation 1698/2005.[2] Under this Regulation rural development policy measures have been clearly divided into four principal areas:

- Increasing the competitiveness of the agriculture and forestry sectors (Axis 1).
- Enhancing the environment and countryside (Axis 2).
- Improving quality of life in rural areas as a whole (Axis 3).
- Assisting locally based projects promoting better quality of life and economic prosperity (Axis 4).

Regulation 1698/2005 adopted a similar template to that which had previously applied under Regulation 1257/99.[3] Member States were required to develop multi-annual rural development plans, for the period January 2008 to December 2013.[4] The Regulation set out twenty-two potential rural development measures and enabled Member States to choose those most suited to their rural areas. Agri-environment and animal welfare measures were exceptions. Member States had an obligation to include, within their rural development plans, measures to promote environmental land management and measures to support animal welfare.

In June 2001, the Göteborg European Council highlighted the important role that rural development policy had to play in helping the European Community

---

1   Council Regulation 1257/99, [1999] OJ L160/80.
2   Council Regulation 1698/2005, [2005] OJ L277/1.
3   *Ibid.* Article 15.
4   *Ibid.* Article 6.

meet its sustainable development objectives.[5] A similar conclusion was reached at the Second European Conference on Rural Development, held in Salzburg in November 2003.[6] Both pointed to the import role that rural development policy should play in ensuring that farmed environments were managed in a way that preserved and enhanced the natural landscape, especially in areas of high nature value. This role was also underlined in the European Commission's proposals for the 2008 CAP Health Check. These highlighted the important role that rural development policy measures would play in helping the European Community to meet new policy objectives in relation to climate change and the development of renewable energy sources.[7] In particular, the European Council's decision in March 2007 to cut carbon dioxide emissions by a minimum of 20 per cent by 2020 and to ensure that bio-fuels contributed a minimum 10 per cent share of petrol and diesel consumption by that date.[8] The 2008 Health proposals also highlighted the role that rural development measures had in relation to existing commitments. They pointed to its role in helping Member States to comply with the Water Framework Directive – under which Member States must act to ensure that objectives on water quality and quantity will be met by 2013.[9] Additionally, it pointed to the major role that agriculture could play in protecting biodiversity, an issue that it identified as being particularly important since the European Community is unlikely to achieve its own objective of halting biodiversity decline by 2010.[10] The Commission's Health Check Proposals ultimately resulted in a legislative requirement that, from 1 January 2010, Member State's Rural Development Programmes should provide

---

5   As set out in the European Commission proposals endorsed by the Council – European Commission, *A Sustainable Europe for a Better World: A European Union Strategy for Sustainable Development*, COM(2001)264 final, 6, 12–13.

6   Information on this European Community sponsored conference can be found at http://ec.europa.eu/agriculture/events/salzburg/index_en.htm [last accessed 19 February 2009]

7   European Commission, Proposal for a Council Regulation amending Regulation 1698/2005 on support for rural development by the European Agricultural Fund for Rural Development, COM(2008) 306 final, 10.

8   The European Council endorsed proposals made in the European Commission's report, *Limiting Global Climate Change to Two Degrees Celsius*, COM (2007) 02 final. See also European Commission, *20 20 by 2020, Europe's Climate Change Opportunity*, COM (2007) 30 final.

9   European Commission, footnote 7 above, 10.

10   *Ibid.* The European Council originally endorsed the European Commission's proposals for a Sustainable Development Strategy as set out in European Commission, *A Sustainable Europe for a Better World: A European Union Strategy for Sustainable Development*, COM (2001) 264 final, 12. This sets out the objective of halting biodiversity loss by 2010. This was also then repeated in the European Community's 6th Environmental Action Plan. See Decision 1600/2002 EC, Decision of the European Parliament and of the Council, laying down the 6th Environmental Action Plan, [2002] OJ L242/14, Article 6(1).

measures which address the Community's priorities in relation to climate change, renewable energies, water management, biodiversity, innovation in these areas and support for the restructuring of the dairy sector.[11]

Axis 1 and Axis 2 of the Community's rural development policy are particularly relevant to the role of that policy in helping the agriculture sector to achieve its environmental objectives. In Axis 1, environmental protection must become an integral part of measures adopted to promote agriculture's competitiveness. In contrast, Axis 2 provides the European Community the opportunity to introduce effective measures to promote positive farm management that protects the environment and enhances rural landscapes.

In Axis 1, measures to integrate environmental protection within agricultural production fall into two main groups: those concerned with farmers' knowledge of environmental protection issues and those providing financial assistance for capital projects that will improve the environmental standards achieved on farms. To help develop farmers' knowledge of environmental issues associated with farming, Regulation 1698/2005 authorises Member States to provide financial support to help reduce costs that farmers would, otherwise, incur in using farm advisory services.[12] The Regulation provides that, at a minimum, these advisory services should provide advice on the statutory management requirements and standards of good agricultural and environmental conditions that have been introduced by the cross-compliance rules and on occupational safety standards created by EC law.[13] The Regulation also enables Member States to provide financial support to young farmers, under the age of 40, who are setting up for the first time as the head of an agricultural holding.[14] Here, it is a pre-condition of such support that these young farmers must 'possess adequate occupational skills and competence' and that they submit a business plan for the development of their farm.[15] It can be implied, clearly, from paragraph 15 of the preamble to the Regulation that, to be eligible, these young farmers will need to have a good awareness of environmental management techniques:

---

11   Council Regulation 1698/2005, [2005] OJ L277/1, Article 16a as inserted by Council Regulation 74/2009, [2009] OJ L30/100.

12   [2005] OJ L277/1, Article 24. The Annex to the Regulation provides that Member States can fund up to 80 per cent of the eligible cost of advisory services, subject to an individual maximum of €1,500 per advice session.

13   *Ibid.* The Community's cross-compliance rules are examined in Chapter 3 of this book.

14   *Ibid.* Article 22.

15   *Ibid.* Article 13 of Commission Regulation 1974/2006/EC, [2006] OJ L368/15, setting down rules for the implementation of the Rural Development Regulation, provides that Member States may grant farmers a period of 36 months from the date of the grant of financial support in which to meet conditions concerning occupational skills and competence.

As regards training, information and diffusion of knowledge, the evolution and specialisation of agriculture and forestry require an appropriate level of technical and economic training, including expertise in new information technologies, as well as an adequate awareness in the fields of product quality, results of research and sustainable management of natural resources, including cross-compliance requirements and the application of production practices compatible with the maintenance and enhancement of landscape and the protection of the environment. It is therefore necessary to broaden the scope of training, information and diffusion of knowledge activities to all adult persons dealing with agricultural, food and forestry matters. These activities cover issues under both the agricultural and forestry competitiveness and the land management and environmental objectives.

In relation to financial support for capital projects, it was previously observed in Chapter 1 that Directive 72/159,[16] had authorised Member States to provide such support for farm development as part of the Community's early structural policy. Today, Regulation 1698/2005 continues to enable Member States to provide financial support for investments to modernise agricultural holdings.[17] This support can be provided for projects that will improve the overall performance of the farm and which respect applicable Community standards.[18] It is clear, from the Community's strategic guidelines, that these investments need not just be aimed at improving the economic performance of a particular farm.[19] They can also include steps taken to provide more environmentally friendly production techniques or to improve the farm's environmental performance. Though, in practice, these two areas need not be mutually exclusive – as, for example, in the case of investments that enable farms to switch to organic production or to the growing energy crops. However, financial support for farm modernisation can only be used to enable farms to meet new Community environmental standards if the farm concerned meets those standards within 36 months of their introduction.[20] In this situation, Regulation 1698/2005 provides a separate measure that enables Member States to contribute towards the costs incurred and income foregone by farmers in meeting new national laws imposing new standards in relation to environmental protection,

---

16   Council Directive 72/159, [1972] OJ L96/1.

17   [2005] OJ L277/1, Article 26.

18   *Ibid.*

19   Council Decision 2006/144/EC on the Community's strategic guidelines for rural development policy (2007–2013), [2006] OJ L55/20/22 as amended by Council Decision 2009/61, [2009] OJ L30/112.

20   See Council Regulation 1698/2005, [2005] OJ L277/1, Article 26 and Commission Regulation 1974/2006, [2006] OJ L368/15, Article 17(1). Though, under Article 17(2) of the Commission Regulation, young farmers setting up as head of their farm may avail of such financial assistance to enable their farm to meet existing Community standards if the investments are undertaken within 36 months of their taking over the farm.

public health, animal and plant health, animal welfare and occupational safety.[21] This support can be provided for a period of up to five years from the introduction of a new law in these areas,[22] provided it is a law that imposes new obligations or restrictions on farming practice, has a significant impact on typical farm operating costs and concerns a significant number of farmers.[23] One example of such a law would be a major extension of Nitrate Vulnerable Zones across all parts of the United Kingdom, following the European Court of Justice's precautionary interpretation of the Nitrates Directive.[24] As a result, livestock farmers, throughout the United Kingdom, have faced the need to make considerable investment to increase manure storage facilities in order to meet the legislative requirements resulting from this extension.

The strategic guidelines state that the 'resources devoted to Axis 2 should contribute to three EU-level priority areas: biodiversity and the preservation and development of high nature value farming and forestry systems and traditional agricultural landscapes; water; climate change.'[25] They encourage Member States to focus their national measures upon a number of key actions:[26]

- Promoting environmental services and animal friendly farming practices
- Preserving the farmed landscape and forests
- Combating climate change
- Consolidating the contribution of organic farming
- Encouraging environmental/economic win-win initiatives [promoting rural tourism and the provision of local amenities]
- Promoting territorial balance [developing rural economies to achieve sustainable balance between urban and rural areas]

This book examines three key measures adopted by the European Community to address these policy goals. This chapter examines the Less Favoured Areas scheme, which seeks to avoid the environmental costs associated with farm abandonment and desertification. Chapter 5 considers the development of the European Community's Agri-Environment scheme and the role it plays in protecting environmental resources. Chapter 6 then examines Community support for organic farming and the environmental contribution made by this sector. The particular importance of the less favoured area and agri-environment measures is illustrated by the fact that they, respectively, account for 14 per cent and

---

21 Council Regulation 1698/2005, *Ibid.* Article 31.
22 Commission Regulation 1974/2006, OJ L368/15, Article 21.
23 Council Regulation 1698/2005, Article 31.
24 See chapter 8 for an examination of this issue.
25 Council Decision 2006/144, footnote 19 above, Annex, paragraph 3.2.
26 *Ibid.* paragraph 3.2.

22 per cent of the Community's entire rural development policy expenditure.[27] Given that Regulation 1698/2005 lists over twenty different rural development measures that Member States can potentially adopt, it is clear that the less favoured areas and agri-environmental measures play a central role in contributing towards the Community's key Axis 2 objectives – enhancing the environment and countryside.

Funding Rural Development Policy Measures

The level of funding available for rural development matters remains a contentious issue. In Chapter 1, reference was made to the fact that Community funding had increased in recent years. In the period up to the mid-1990s, Community funding for agricultural structural policy never exceeded 5 per cent of its agricultural expenditure.[28] In contrast, by 2002, Community spending on rural development policy measures accounted for 16 per cent of its agricultural expenditure.[29] Despite this, the European Commission's proposals for the 2003 Mid-Term Review reform of the Common Agricultural Policy made it clear that rural development policy measures were still inadequately resourced.[30] The European Commission proposed 'modulation' as a method of channelling additional Community funds into rural development policy.

The practice of modulation refers to deductions being made from the direct payments made to farmers as part of agricultural production policy, with the monies saved being channelled into rural development measures. It was, originally, introduced as a discretionary measure under Regulation 1259/99.[31] This Regulation gave Member States the discretion to reduce the value of direct payments by up to 20 per cent. This discretion could be exercised where any of the following circumstances existed:

1. the farm labour force fell below limits set by the Member State
2. the overall prosperity of farm holdings exceeded limits set by the Member State
3. the total amount of payments made under Common Agricultural Policy support schemes exceeded limits set by the Member State

---

27   See European Commission (2007), *Rural Development in the European Union: Statistical and Economic Information* (Brussels: European Commission Directorate General for Agriculture) 20.

28   Grant, W. (1997), *The Common Agricultural Policy* (London: Macmillan) 71.

29   European Commission, *Mid Term Review of the Common Agricultural Policy*, COM (2002)394, 9.

30   *Ibid.*

31   [1999] OJ L160/113, Article 4.

The United Kingdom acted upon the discretion provided, introducing deductions of 2.5 per cent in 2001, rising to 3 per cent in 2002 and 3.5 per cent in 2003.[32] Elsewhere in the European Community, however, only France and Portugal also chose to adopt such action.[33]

The rate of modulation within the United Kingdom was due to rise to 4.5 per cent in 2005. However, before this could happen, the European Community replaced discretionary modulation with a compulsory scheme. Under Regulation 1782/2003 modulation would affect all farmers who received €5,000 or more in direct payments through the operation of the common organisations.[34] A deduction of 3 per cent was made from these payments in 2005, rising to 4 per cent in 2006 and reaching a ceiling of 5 per cent in the period 2007–2012.[35] A proposal that a maximum ceiling of €300,000 per annum should be set on the amount of direct aid that could be made to any one claimant was not incorporated in the Regulation.[36] The monies saved through compulsory modulation were to be made available to Member States as additional Community support for rural development measures. The actual amount allocated to each Member State varied according to criteria such as the size of their agricultural area, their levels of agricultural employment and their gross domestic product.[37] However, each Member State was guaranteed to receive at least 80 per cent of the funds that had been deducted from its farmers.[38]

The 2008 Health CAP Check sought to promote a wider role for rural development measures, both in terms of helping the Community to meet new environmental objectives in relation to climate change and energy policy and in supporting existing policies on sustainable water management and protecting biodiversity.[39] The Community's budgetary ceilings for agriculture are already set until 2013; provisions in the Health Check, therefore, mean that greater financial support for rural development should come via a percentage increase in direct payments deducted through compulsory modulation.[40] Modulation rates will increase to 7 per cent in 2009 and then by an additional 1 per cent per annum between 2009 and 2012, with an additional 4 per cent reduction also being

---

32   See the Common Agricultural Policy Support Schemes (Modulation) Regulations 2000, SI2000/3127.

33   See Cardwell, M. (2004), *The European Model of Agriculture* (Oxford: Oxford University Press) 198.

34   [2003] OJ L270/1, Articles 10 and 12.

35   Article 10(1).

36   See European Commission, footnote 7 above, 23.

37   Regulation 1782/2003, Article 10(3).

38   *Ibid.*

39   European Commission, footnote 7 above, 10.

40   The Financial framework for EU budgetary expenditure between 2007 and 2013 was set by the Inter-Institutional Agreement on budgetary discipline and sound financial management adopted by the European Parliament, the Council of the European Union and the European Commission on 17th May 2006, as amended by decision 2008/29/EC of the European Parliament and of the Council, [2008] OJ L6/7.

imposed, in each of these years, on farmers whose total direct payments exceed €300,000.[41] The thresholds and the percentage rates of modulation are set out in Table 4.1 below:[42]

Table 4.1   Compulsory Modulation Rates 2009–2012

Thresholds (Euros)	2009	2010	2011	2012
< 5,000	0	0	0	0
5,000–99,000	7	8	9	10
300,000 +	11	12	13	14

In addition to compulsory modulation, the European Community has allowed Member States to choose to operate a further voluntary modulation scheme, again with the objective of raising funds for their rural development measures. Member States can elect to deduct up to 20 per cent of farmers' direct payments for this purpose over the period 2007 to 2012.[43] The United Kingdom has chosen to avail itself of this option, though the devolved governments have introduced differing voluntary modulation rates.[44]

One further, important, influence upon the resources available for rural development measures is the fact that the European Community continues, only, to part-fund these measures. As noted in Chapter 1, the Community only provided part-funding for agricultural structural policy measures and, subsequently, continued

---

41   Council Regulation 73/2009, [2009] OJ L30/16, Article 7.

42   Under the European Commission's proposals, modulation rates would have depended upon the level of direct payment received by farmers. Four categories would have existed, €5,000–€99,000, €100,000–€199,000, €200,000–€299,000 and €300,000+. Higher modulation levels would have been set in each successive category and would have continued to increase annually between now and 2012. Farmers receiving more than €300,000 per annum in direct payments would have faced a 22 per cent deduction by 2012. See European Commission, footnote 7 above, 8.

43   Under Council Regulation 1782/2007, [2007] OJ L95/1 laying down rules for voluntary modulation of direct payments.

44   In England voluntary modulation rates have been set at 12 per cent for 2007, 13 per cent for 2008 and 14 per cent in 2009–2012. In Scotland the voluntary modulation rate was 5 per cent in 2007 and 8 per cent in 2008. This will rise to 8.5 per cent in 2009 and 9 per cent in 2010. Wales had a voluntary modulation rate of 0 per cent in 2007, but introduced deductions of 2.5 per cent in 2008 and 4.2 per cent in 2009, rising to 5.8 per cent in 2011 and 6.5 per cent in 2012. In Northern Ireland a voluntary modulation rate of 5 per cent was introduced in 2007, which will rise to 9 per cent in 2011. See www.defra.gov.uk/rural/rdpe/secta.htm#q1 [last accessed 19 February, 2009].

this policy with rural development policy measures. Initially, when agricultural structural policy measures were first introduced, the Community refunded up to 25 per cent of eligible national spending upon these measures. Today, the level of the Community contribution has increased. It will now reimburse up to 60 per cent of eligible national spending on Axis 2 measures, and, indeed, up to 85 per cent of that expenditure in convergence areas (regions whose gross domestic product is less than 75 per cent of the EU average).[45] One consequence of the continuation of part-funding is that the breadth and depth of the rural development plans, put in place within each Member State, will depend heavily upon the willingness of national treasuries to provide additional funding to bolster that available from the Community.

Less Favoured Areas

The Less Favoured Areas scheme is one of the most widely adopted measures available within the Community's Rural Development Policy, with 56 per cent of agricultural land across the Community designated as being less favoured.[46] The origins of the scheme can be traced back to agricultural policy within the United Kingdom following World War Two. This policy provided supplementary payments to farmers in hilly and mountainous areas, to compensate them for the poor agricultural conditions that they experienced there. When the United Kingdom joined the European Community, it negotiated the introduction of a similar scheme as part of the Common Agricultural Policy. This was introduced by Directive 75/268 on mountain and hill farming in agriculturally less favoured areas.[47] According to Article 1 of this Directive, its objective was 'the continuation of farming, thereby maintaining a minimum population level, or conserving the countryside'.

Directive 75/268 was based upon recognition that poor agricultural land, in hilly and mountainous areas, limited the earning capacity of those farmers based there and increased the risk that they would abandon their land, thereby threatening the sustainability of rural communities and environmental degradation upon the untended land. The Directive required Member States to nominate the areas within which it would apply – these nominations then being subject to the approval of the Council of Ministers. Within these 'less favoured areas' Member States were authorised to make direct income payments to farmers or to provide

45   See Council Regulation 1783/2003 amending Council Regulation 1257/99 on support for rural development from the European Agricultural Guidance and Guarantee Fund [2003] OJ L270/70.

46   See Court of Auditors, *Special Report No.4/2003 concerning rural development: support for less favoured areas*, [2003] OJ C151/1, 4.

47   Council Directive 75/268 [1975] OJ L128/1.

additional financial grants for farm development or diversification into tourist or craft activities.

The initial less favoured area policy was, essentially, a social measure designed to prevent rural depopulation. In this regard it was a qualified success. Farm numbers have fallen steadily in all Member States, since the end of the Second World War, as have employment levels within agriculture. The Less Favoured Areas policy has, however, helped to slow the rate of decline. In terms of environmental protection, the Directive was the first piece of Community legislation to highlight the environmental role played by farming. However, as Fennell has pointed out, the environmental aspects of the Directive were 'largely incidental.'[48] In practice, the Directive's environmental credentials were limited. It operated on the basis that maintaining a farmed landscape would ensure the conservation of the countryside, preventing the gradual reclamation of deserted farmlands, by shrub and forest, that would obliterate other habitats and landscapes. One of the principal mechanisms adopted by the Less Favoured Area scheme, to achieve this objective, was the use of additional payments to farmers based upon their livestock numbers.[49] However, this, in itself, became a source of environmental damage, as farmers increased their livestock numbers in order to maximise their payments. A number of studies have shown that this resulted in environmental damage through overgrazing.[50] Ironically, this damage was compounded by the rules underpinning the operation of the common organisation of the market in sheep meat. Regulation 3013/89 provided for sheep annual premium payments to be made, in full, on up to 1,000 sheep per farmer within less favoured areas, but on only 500 sheep in other areas.[51] Farmers holding sheep in excess of these numbers received only 50 per cent of the sheep annual premium payment on additional sheep. However, by definition, less favoured areas were usually less able to support higher sheep numbers without environmental damage being caused. This anomaly was removed in 1994,[52] but its existence serves to illustrate that early measures were principally driven by

---

48   Fennell, R. (1997), *The Common Agricultural Policy: Continuity and Change* (Oxford: Clarendon Press) 345.

49   See Directive 75/268/EEC, Article 3, or subsequently Council Regulation 950/97 ([1997] OJ L142/1) Article 19.

50   See, for example, Jenkins, T.N. (1990), *Future Harvests, The Economics of Farming and the Environment: Proposals for Action* (London: Council for the Protection of Rural England) 29; Wathern, P. Less Favoured Areas and Environmentally Sensitive Areas: A European Dimension to the Rural Environment, in Howarth, W., Rodgers, C. (eds) (1995), *Agriculture, Conservation and Land Use: Law and Policy Issues for Rural Areas* (Cardiff: University of Wales Press) 206, The Wildlife Trusts (1996), *Crisis in the Hills: Overgrazing in the Uplands* (Lincoln: The Wildlife Trusts) 10; Wildlife and Countryside Link (1997), *Farming in the Uplands in the Next Millennium* (London: Wildlife and Countryside Link) 5; Lovegrove, R., Shrubb, M. and Williams, I. (1995), *Silent Fields – The Current Status of Farmland Birds in Wales* (Sandy: Royal Society for the Protection of Birds).

51   Council Regulation 3013/89, [1989] OJ L289/1.

52   By Council Regulation 233/94, [1994] OJ L30/9.

social considerations. In truth, little thought had been given to the environmental implications of such policy measures.

It also became increasingly clear that initial European Community policy, within agriculturally less favoured areas, had also ignored one vital factor: the environmental value of many areas of farmland situated within less favoured areas. A number of publications have illustrated that areas of high natural value farm land tend to be found within agriculturally less favourable areas.[53] This is the result of restrictions placed upon agriculture by the natural conditions found in many such areas, which discouraged agricultural intensification. Consequently, most areas of low intensity farming are often found within less favoured areas. For example, a study of low intensity farmland in the United Kingdom showed that 89 per cent of this land was located within less favoured areas.[54] In turn, as Baldock and Beaufoy explain, these areas of low intensity farming often contain areas of high natural value:

High Natural Value (HNV) farming systems are predominantly low-intensity systems which often involve a relatively complex interrelationship with the natural environment. They maintain important habitats both on the cultivated or grazed area (for example, cereal steppes and semi-natural grasslands) and in features such as hedgerows, ponds and trees, which, historically, were integrated with the farming system. 'Modern' farming, by contrast, tends to be intensive in its exploitation of the land and in its use of inputs, excluding all but the most common forms of wildlife from the productive area, as well as highly rationalised in its exploitation of the land, neglecting or removing features which no longer serve a productive function. The semi-natural habitats currently maintained by HNV farming are particularly important for nature conservation in the EC because of the almost total disappearance of large-scale natural habitats.

High natural value farmland retained semi-natural habitats and supported a broad range of wildlife. However, in practice, less favoured area payments

---

53   See Baldock, D., Beaufoy, G. (1993), *Nature Conservation and New Directions in the EC Common Agricultural Policy* (London: Institute for European Environmental Policy); Beaufoy, G., Baldock, D., Clark, J. (1994), *The Nature of Farming: Low Intensity Farming Systems in Nine European Countries* (London: Institute for European Environmental Policy), Hellegers, P., Godeschalk, F. (1998), *Farming in High Nature Value Regions: The Role of Agricultural Policy in Maintaining HNV Farming Systems in Europe* (The Hague: Agricultural Economics Research Institute), Dax, T., Hellegers, P. *Policies for Less Favoured Areas*, in Brouwer, F. and Lowe, P. (2000), *CAP Regimes and the European Countryside* (Wallingford: CAB International), Cooper, T., et al. (2006), *An Evaluation of the Less Favoured Area Measure in the 25 Member States of the European Union* (London: Institute for European Environmental Policy), chapter 9.

54   Wilson, J. The Extent and Distribution of Low Intensity Agricultural Land in Britain, in Joint Nature Conservation Committee (JNCC) (1991), *Birds and Pastoral Agriculture in Europe* (Peterborough: JNCC).

recognised only the lower agricultural productivity of such areas, taking no account of their environmental value or of the farmers' role in managing them. This situation is best summed up by the following general criticism, made by Bignal and McCracken:[55]

> It is ironic that many environmental initiatives on farmland tend to concentrate (often with little prospect of success) on reversing actions that have been destructive, yet tend to ignore practices that are currently benign and could be sustained. In this respect, most environmental initiatives tend to reward some farmers for their previously destructive activities but not others for their contribution to the maintenance of biodiversity.

One exception did exist, in the Netherlands. There, Directive 75/268 was interpreted liberally in order to promote the continuation of traditional farming within wetland areas and to discourage farmers from draining these areas.[56] The Dutch government chose to restrict the application of the Directive to recognised wetland areas and used it to help fund management agreements under which farmers agreed to retain existing wetland areas on their farms.[57] In the United Kingdom, however, government officials concluded that the Dutch approach was beyond the scope of the Directive. Consequently, when discussion arose on the need to protect the traditional wetland landscape of the Norfolk Broads, the United Kingdom government lobbied the European Community to introduce a new, environmental, scheme.[58] As noted in Chapter 6, this was how the Environmentally Sensitive Areas scheme came into being.

Greening Less Favoured Area Policy

Since its introduction in the Common Agricultural Policy in Directive 75/268, the European Community's less favoured area scheme has remained an important element of that policy. Initially, Directive 75/268 was repealed and replaced by Regulation 797/85,[59] which itself was replaced by Regulation 2328/91.[60] More

---

55   Bignal, E., McCracken, D. The Ecological Resources of European Farmland, in Whitby, M. (ed.) (1996), *The European Environment and CAP Reform: Policies and Prospects for Conservation* (Wallingford: CAB International) 30.

56   Baldock, D., Lowe, P. The Development of European Agri-Environment Policy, in Whitby, M. (ed.) (1996), *The European Environment and CAP Reform: Policies and Prospects for Conservation* (Wallingford: CAB International) 14.

57   See Baldock, D., Beaufoy, G., footnote 53 above, 61 and Potter, C. (1998), *Against the Grain: Agri-Environmental Reform in the United States and the European Union* (Wallingford: CAB International) 107.

58   Baldock, D. and Lowe, P., footnote 56 above, 14.

59   Council Regulation 797/85[1985] OJ L93/1.

60   Council Regulation 2328/91 [1991] OJ L218/1.

recently, the less favoured area scheme has formed part of the rural development policy introduced by Regulation 1257/99[61] and continued by Regulation 1698/2005.[62] These subsequent Council Regulations have also put in place measures designed to modernise the less favoured area scheme, in particular by promoting its environmental credentials.

Regulation 2328/91 introduced a requirement that less favoured area payments should only be available to farmers whose stocking densities did not exceed 1.4 livestock units per hectare.[63] This had an incidental environmental effect, in discouraging increased livestock numbers. However, its primary goal was to avoid the impact that the additional sheep annual premium payments would have had upon the Community budget. Despite criticisms of their environmental impact, less favoured area payments continued to be calculated on the basis of the number of livestock that each farmer owned. From an environmental perspective, the new measures also suffered from the fact that it is extremely difficult to establish effective uniform measures in Brussels. For example, a stocking density of 1.4 livestock units per hectare was likely to prove to be too high to prevent overgrazing occurring in many upland areas. A danger also existed in that it could be regarded as a target that individual farmers should meet to maximise their potential income from less favoured area payments. Such environmental criticisms were, finally, acknowledged by the European Community in 1999, when Regulation 1257/99 abolished the practice of calculating less favoured area payments on the basis of the number of livestock units each farmer owned.[64] Instead, Member States were required to calculate the payments on the basis of the land area farmed by individual farmers. This remains the basis of the scheme today.[65]

Regulation 2328/91 also gave Member States discretion to introduce cross-compliance measures. They were authorised to attach conditions to less favoured area payments that encouraged farmers to adopt practices that were compatible with the protection of the environment and the protection of the countryside.[66] This discretion was, subsequently, replaced by an obligation to attach environmental conditions to these payments. Firstly, Regulation 1257/99 required farmers to 'apply usual good farming practices compatible with the need to safeguard the environment and maintain the countryside, in particular by sustainable farming.'[67] This approach, however, was criticised by the Court of Auditors. They noted that the Regulation provided no verifiable, clear or consistent definition of 'good agricultural practices', with the result that Member State compliance was difficult

---

61   Council Regulation 1257/99, [1997] OJ L160/80.
62   Council Regulation 1698/2005 [2005] OJ L277/1.
63   See Article 19(1) (a).
64   Council Regulation 1257/99, Article 14.
65   Council Regulation 1698/2005, Article 37.
66   Council Regulation 2328/1991, Article 18(3).
67   Council Regulation 1257/99, Article 14(2).

to verify.[68] More recently, Regulation 1698/2005 replaced these cross-compliance measures and required farmers receiving less favoured area payments to comply with the same cross-compliance obligations as farmers receiving direct payments such as the Single Farm Payment.[69] Consequently, these farmers must now abide by the national laws implementing the particular European Community legislation identified in Chapter 3, as well as complying with national codes of good agricultural and environmental condition.[70] As explained in Chapter 3, these codes must now address specific agricultural and environmental issues and standards. Failure to comply with these obligations will lead to the level of less favoured area payment being reduced, whilst repeated breaches may lead to farmers being excluded from eligibility altogether.

Identifying Less Favoured Areas

Although the less favoured area scheme was initially developed, principally, as a social measure, the European Community now seeks to place greater emphasis on the environmental objectives of the scheme. This is evident from its inclusion in Axis 2 of the European Community's rural development policy, which is entitled 'Improving the environment and the countryside.' Additionally, Regulation 1698/2005 specifically stipulates that the modern objective of the scheme is to target the 'sustainable use of agricultural land'.[71] The Regulation established that the areas eligible for payments fall into two main categories: mountain areas affected by significant natural handicaps and other, non-mountain, areas affected by handicaps. However, in practice, the Regulation then divides this latter category into two sub-categories: (i) non-mountain areas affected by significant natural handicaps and (ii) non-mountain areas affected by specific handicaps. In 2003, the Court of Auditors published a highly critical report which highlighted, inter alia, that the European Commission had insufficient evidence that existing less favoured area classifications remained valid, or any clear evaluation of the impact of the scheme.[72] In the light of this criticism the Council of the European Union asked the Commission to review Member State implementation of the scheme and to bring forward new proposals for a future less favoured area scheme and for future less favoured area designations. The present scheme and designations will remain in place until 2010, pending the introduction of this new scheme.

---

68   Court of Auditors, footnote 46 above, 12. For a detailed examination of the provisions implemented by individual Member States see Cooper, T., et al. (2006), *An Evaluation of the Less Favoured Area Measure Scheme in the 25 Member States of the European Union* (London: Institute of European Environmental Policy).

69   Council Regulation 1698/2005, Article 51.

70   *Ibid.* As amended by Council Regulation 74/2009, [2009] OJ L300/100.

71   Council Regulation 1698/2005, Article 36.

72   Court of Auditors, footnote 44 above, 6.

Mountain Areas

These are areas that become eligible for less favoured area payments as a result of the impact that natural conditions have upon agricultural activity – the fact that those conditions result in a 'considerable limitation of the possibilities of using the land and an appreciable increase in the cost of working it...'[73] This can be because of, either, the climatic impact of altitude, in substantially shortening the growing season or, at, lower altitudes, the limitations imposed by large areas of steep slopes, which prevent the use of machinery and/or require expensive special equipment.[74] In addition, the Regulation stipulates that areas north of the 62nd parallel and certain adjacent areas will also be considered to be mountain areas. These criteria mirror those that applied in relation to mountain areas under the less favoured area scheme that previously operated, under Regulation 1257/99. Under that scheme they accounted for some 35 per cent of the less favoured areas identified by Member States.[75]

Areas Affected by Significant Natural Handicaps

Previously, in the less favoured area scheme offered under Regulation 1257/99, Member States were able to identify non-mountain lands as being agriculturally less favourable if they could show that farms in those areas were at risk of abandonment and that conservation of the countryside was necessary.[76] Risk of abandonment was determined by identifying areas with poor land productivity, below average agricultural production and a low or dwindling population that was predominantly dependent upon agriculture.[77] These lands accounted for some 61 per cent of all less favoured areas that received support under Council Regulations.[78] The Court of Auditors report, however, suggested that many of these designations were based on outdated socio-economic data.[79] One reaction to this can be seen in the fact that Regulation 1698/2005 actually removed the socio-economic requirement that eligible areas had to have a low or dwindling population that was dependent on agriculture. Instead, the Regulation concentrates on the natural handicaps that exist in non-mountain areas. These areas are now eligible for inclusion if they are affected by significant natural handicaps, principally low soil fertility or poor climatic conditions, and it can also be demonstrated that 'maintaining extensive

---

73   Council Regulation 1698/2005, Article 50(2).

74   *Ibid.* A combination of these factors can be taken into account where, under Regulation 50(2) (b) 'the handicap resulting from each taken separately is less acute but the combination of the two gives rise to an equivalent handicap.'

75   Court of Auditors, footnote 46 above, 6.

76   Council Regulation 1257/99, Article 19.

77   *Ibid.*

78   Court of Auditors, footnote 46 above, 6.

79   *Ibid.* 15–16.

farming activity is important for the management of the land'.[80] Looking to the future, this less favoured area category is likely to be at the forefront of future reform of the scheme. Certainly, the Commission's public consultation on the reform of the less favoured area scheme principally concentrated upon options for reform of the category.[81]

Areas Affected by Specific Other Handicaps

Non-mountain areas affected by other 'specific handicaps' are also eligible for designation as less favoured areas, where it can be shown that the continuation of farming is important to 'conserve or improve the environment, maintain the countryside and preserve the tourist potential of the area or in order to protect the coastline.'[82] However, Member States are only able to identify up 10 per cent of their territory as being less favoured land on account of specific non-natural handicaps.[83] Previously, some 4 per cent of the agricultural land identified as being less favourable under Regulation 1257/99 belonged to this category.[84]

Natura 2000 Payments

Regulation 1257/99 also extended the scope of the less favoured area scheme even further, in environmental terms. It gave Member States discretion to provide less favoured area payments to farmers who face restrictions upon the agricultural use of their land as a result of Community environmental protection legislation.[85] This enabled Member States to compensate farmers for costs they incurred or profits they lost as a result of this legislation. These payments were, primarily, concerned with the impact of the Wild Birds Directive 1979 and the Habitats Directive 1992, under which the designation of farmland as a Special Protection Areas or as a Special Areas of Conservation would place limitations upon farmers' use of that land.[86] Additionally, the payments could also compensate farmers for restrictions imposed under the Water Framework Directive 2000.[87]

Under Regulation 1698/2005 these payments have evolved into a stand-alone scheme 'the Natura 2000 payment scheme', which now operates separately from the less favoured area scheme.[88] Initially, these payments could only be made on a

---

80   Council Regulation 1698/2005, Article 50(3).

81   European Commission (2008), *Review of the 'Less Favoured Area' Scheme: Public Consultation Document for Impact Assessment* (Brussels: European Commission).

82   Council Regulation 1698/2005, Article 50(3).

83   *Ibid.*

84   Court of Auditors, footnote 44 above, 6.

85   Council Regulation 1257/99, Article 16.

86   See Chapter 7 for further details of the operation of these Directives.

87   See Chapter 8 for further details of the operation of this Directive.

88   Council Regulation 1698/2005, Article 38.

maximum of 10 per cent of each Member State's agricultural land area.[89] However, this restriction was removed in Regulation 1698/2005.

An Evaluation of the Less Favoured Area Scheme

Recent reforms to the Community's less favoured area policy have played an important role in reducing the environmental side-effects associated with earlier less favoured area policy measures. In particular, the introduction of area based payments and of compulsory cross-compliance has been particularly important. However, these measures really only ensure that farmers receiving less favoured area payments avoid causing environmental damage and observe basic environmental conditions. Arguably, this is not enough for a scheme which forms part of Axis 2 of the Community's rural development policy – with its title 'improving the environment and the countryside'. Certainly, the scheme does continue to make an important environmental contribution by supporting extensive farming in areas that might, otherwise, be subject to desertification. However, the scheme remains predicated upon the need to compensate farmers for the production difficulties that they face as a result of physical constraints upon their farms. This is evident in the fact that Regulation 1698/2005's instruction, that payments should 'compensate for farmers' additional costs and income foregone related to the handicap for agricultural production in the area concerned.'[90] This has been the goal of the scheme since it began in 1975, when this coalesced with the production orientation of the Community's agricultural production policies. At that time, little thought had been given to the environmental value of these areas. Today, as noted earlier, most areas of high natural value farmland are found within less favoured areas. However, the Community's less favoured areas scheme does little to encourage positive management of such land. For example, the compulsory standards of good agricultural and environmental condition, which form part of the Community's cross-compliance mechanism, only require farmers receiving less favoured area payments to observe basic management requirements. Within the United Kingdom, recent policy measures have shown a desire to use the Community's less favoured areas scheme to provide greater support for environmental management. Wales, for example, operates the Tir Mynydd scheme within less favoured areas.[91] Farmers there can qualify for a payment enhancement of 10 per cent if they satisfy one of several criteria and for an enhancement of 20 per cent if they satisfy two or more of these criteria.[92] The criteria include maintaining stocking densities of no more than 1.2 livestock units per hectare, having a farm

---

89   Council Regulation 1257/99, Article 21.

90   Council Regulation 1698/2005, Article 37(1).

91   Introduced by the Tir Mynydd (Wales) Regulations 2001 (SI 2001/496. W.23) as amended by SI 2001/154 (W.61), SI 2002/1806 (W.176) and SI 2005/1269 (W.89).

92   The Tir Mynydd (Wales) Regulations 2001, as amended, Regulation 7.

that has converted to organic status (so long as the farmer is not already receiving any other organic aids) or exercising grazing rights on a common[93] and agreeing with all other graziers to remove all stock from that land for a period of 3 months between September and February each year.[94] In England, recent proposals go even further. DEFRA has proposed that England's existing less favoured area scheme, the Hill Farming Allowance,[95] should be replaced by an Uplands Entry Level Stewardship scheme. This would, effectively, amalgamate the Community's less favoured areas and agri-environment schemes together within one national scheme operating in England.[96] Farmers would enter a management agreement with DEFRA and, in return for receiving payments, would undertake to manage their land in order to preserve and enhanced its environmental and landscape value. A similar amalgamation of both schemes, at European Community, level has the potential to provide more funding for positive environmental management. To be effective, such a reform would also need to be accompanied by a reconsideration of the areas classified under the amalgamated schemes. Instead of identifying areas in which physical handicaps restricted agricultural production, there would need to be a move towards identifying areas on the basis of their high nature value. As previously noted, the Court of Auditors, has challenged the validity of the existing less favoured area designations, made on the basis of handicaps upon agricultural production. Equally, it is also true that, whilst most areas of high nature value are found within areas presently designated as being agriculturally less favoured, the present classification system does not provide an accurate reflection of farmland's environmental impact. Currently, 56 per cent of the Community's agricultural land is classified as being agriculturally less favourable.[97] In contrast only 15 to 25 per cent of that land is estimated to contain high nature value farmland.[98] A revised system, based upon the identification of these areas of high nature value, could, for example, provide for farmers in these areas to receive enhanced payments in return for managing this land. Alternatively, as long as it does not promote positive environmental management, the Community's less favoured areas scheme will remain, as Fennell suggested, a rather blunt instrument. This also leaves the task of funding positive environmental management to the separate European Community agri-environment land management scheme, which is examined in Chapter 5.

---

93 Under the Commons Registration Act 1965.
94 The Tir Mynydd (Wales) Regulations 2001 as amended, Regulation 8.
95 Under the Hill Farm Allowance Regulations 2008 (SI 2008/51).
96 Agri-Environment schemes are examined in Chapter 5.
97 Court of Auditors, footnote 44 above, 3.
98 European Environment Agency, (2004), *High Nature Value Farmland: Characteristics, Trends and Policy Challenges* (Copenhagen: European Environment Agency) 16.

Chapter 5

Agri-Environment Policy

The Role Played by Agri-Environment Agreements

Agri-environment agreements were introduced into the Common Agricultural Policy in the mid-1980s. They have, since, become a vital part of the European Community's strategy to integrate environmental protection within agricultural policy. The European Commission has explained this strategy in the following terms:[1]

> The philosophy underpinning the environmental aspects of the CAP reform
> is that farmers should be expected to observe basic environmental standards
> without compensation. However, wherever society desires that farmers deliver
> an environmental service beyond this base-line level, this service should be
> specifically purchased through agri-environmental measures.

The first part of this philosophy has been addressed through attaching cross-compliance requirements to the direct payments made within CAP production policy.[2] In contrast, agri-environment agreements play a key role in delivering the second aspect – rewarding farmers who provide an environmental service that goes beyond the observation of basic environmental standards.

Agri-environment agreements provide a mechanism through which national agriculture authorities can offer farmers voluntary management contracts. Participating farmers enter a management contract under which they agree to manage their land for a fixed number of years in a way that will develop its environmental value, in return for receiving payments. The scheme, therefore, provides a policy measure through which Member States can seek to protect and maintain existing environmental features and to restore and enhance those features. It also provides an opportunity to tailor agri-environment measures, to target environmental issues of particular importance in individual regions – such as the conservation of wildlife species and habitats or the protection of traditional landscapes and places of recreation.[3]

---

1   European Commission, *Directions Towards Sustainable Agriculture*, COM (99) 22 Final, 28.

2   See Chapter 3 for further details.

3   Wildlife Link (1995), *Agri-Environment Management Agreements: Their Benefits and Future* (London: Wildlife Link) 6.

The introduction of agri-environment agreements filled a lacuna within the CAP. Most highly valued rural landscapes are the product of agricultural management systems that created and preserved them. These management systems also created and preserved important wildlife habitats. However, as was outlined in Chapter 2, the initial Common Agricultural Policy only rewarded agricultural production. No public market existed to pay for the production of conservation goods. In this situation, also highlighted in Chapter 2, many valued landscapes and important habitats were damaged or destroyed as farmers intensified production in response to the production signals provided by the Common Agricultural Policy. The development of agri-environment agreements provided an opportunity for the Community to both recognise the role played by farmers in environmental management and to protect the rural environment.[4] Additionally, on poorer agricultural land, it supplemented the less favoured area scheme, providing support for traditional farming practices in areas within which environmental damage was threatened by the risk of farm abandonment.[5]

The development of agri-environment agreements fitted well with the pressures for agricultural policy reform. At a time when the European Community was under pressure to tackle surplus production and its cost to the European Community budget, the scheme discouraged participating farmers from intensifying their production.[6]

The environmental payments, provided under agri-environment agreements, also gave farmers an additional income source. As the European Commission has observed:[7]

> If conservation is to become a major goal of farm management…., it must be recognised that this is likely to entail limitations on land improvement, a preference for extensive methods and the performance of tasks which may no longer be customary on modern farms. In general, this approach is likely to prevent a farmer from exploiting the maximum earning potential of the farm. Hence the argument for compensation. Equally, conservation may require the maintenance of traditional systems which would otherwise be abandoned, so the question of incentives arises.

---

4   See, for example, European Commission (1986), *Agriculture and Environment: Management Agreements in Four Countries of the European Community* (Luxembourg: Office for Official Publications of the European Communities) 15.

5   *Ibid.*

6   See, for example, Whitby, M., Lowe, P. (1995), The Political and Economic Roots of Environmental Policy in Agriculture, in Whitby, M. (ed.) (1995), *Incentives for Countryside Management: The Case of Environmentally Sensitive Areas* (Wallingford: CAB International) 10.

7   European Commission, footnote 4 above, 10.

Ultimately, payments made under agri-environment agreements can provide for several eventualities.[8] They can support the continuation of extensive farming methods, prevent farmers from moving to more intensive methods and discourage them from abandoning their land. They reward farmers for positive management that enhances the environmental value of the farm and they also compensate farmers for accepting restrictions upon their farming practice. Additionally, agri-environment agreements operate alongside more traditional regulatory methods which, either through cross-compliance or the operation of European Community environmental directives, punish farmers who fail to achieve basic environmental standards.

The Development of Agri-Environment Agreements: From the Norfolk Broads to Brussels

The European Community's current agri-environment scheme had its roots in the conservation problems experienced in the Norfolk Broads in the early 1980s. Farming there had traditionally been livestock based and supported a wetland marsh landscape. However, in the face of falling livestock profits, farmers had been ploughing up the grassland, draining the land and planting, more profitable, arable crops. In an attempt to protect the traditional landscape, the Broads Grazing Marshes Conservation Scheme was introduced in January 1985.[9] The scheme was jointly administered by the Countryside Commission and the Ministry of Agriculture, Fisheries and Food (MAFF).

The Broads Grazing Marshes Conservation Scheme gave farmers the opportunity to enter a three year management agreement. Under this agreement they received annual payments of £125 per hectare in return for observing the scheme's conditions. These required farmers to keep their land as permanent grassland and to observe limits on minimum and maximum stocking densities and on their use of fertilisers and pesticides. As Potter has noted, the scheme was quite revolutionary.[10]

It was open to all farmers within a specific area and provided participating farmers with flat rate payments, regardless of whether they had shown any

---

8   *Ibid.* 34.

9   For further details of this scheme see variously Turner, K., The Broads Grazing Marshes Conservation Scheme, [1985] 10 *Landscape Research* 28–29; O'Riordan, T., Halvergate: The Policy and Politics of Change, in Gilg. A., (ed.) (1985), *Countryside Planning Yearbook* (Norwich: Geobooks); or Baldock, D., Lowe, P. (1996), The Development of European Agri-Environment Policy, in M. Whitby (ed.) (1996), *The European Environment and CAP Reform: Policies and Prospects for Conservation* (Wallingford: CAB International).

10   Potter, C. (1998), *Against the Grain: Agri-Environmental Reform in the United States and the European Union* (Wallingford: CAB International) 49.

intention of ploughing and draining their fields. In contrast, the United Kingdom's national nature conservation designation, the Site of Special Scientific Interest ('SSSI'), then only made payments to farmers who indicated a creditable intention of conducting works that would damage or destroy the special interest of their land. Payments under the SSSI scheme were also individually calculated on the basis of the profit each farmer would forego if that work were not conducted. Potter noted that adopting this approach in the Broads Grazing Marshes Conservation Scheme would have placed the scheme in competition with CAP production payments. High levels of compensation would have been needed to compensate farmers for not receiving the agricultural subsidies they could have received if they had planted and harvested arable crops that were already in surplus.[11]

The principal aims of the Broads Grazing Marshes Conservation Scheme were to encourage the continuation of traditional livestock grazing practices, thereby conserving the unique character of the landscape, and to consider whether the scheme could also be applied successfully in other areas.[12] The scheme proved very successful, with 8,000 hectares of permanent grassland, out of 9,500 eligible hectares, becoming enrolled.[13]

The timing of the introduction of the Broads Grazing Marshes Conservation Scheme corresponded with the European Commission's of proposals for a new regulation to replace the four agricultural structures directives adopted by the Community in the 1970s.[14] Its initial proposals were criticised by the House of Lords Select Committee on the European Communities, as failing to include measures to promote diversification and environmental enhancement.[15] Equally, MAFF was also concerned that it did not actually have power, under EC Law, to implement its proposals for the Broads Grazing Marshes Conservation Scheme.[16] This persuaded MAFF to lobby the Commission to ask it to make provision for measures 'to encourage farming practices which are consonent with conservation' within areas of recognised environmental value.[17]

MAFF's lobbying was, ultimately, successful and the European Commission's proposals subsequently became Regulation 797/85 on improving the efficiency of

---

11   *Ibid.*

12   Turner, K., footnote 9 above, 28.

13   *Ibid.* 29.

14   As noted in Chapter 1, the Community had introduced Directive 72/159 on farm modernisation, Directive 72/160 on early retirement from farming and Directive 72/161 on socio-economic guidance for farmers and subsequently also introduced Directive 75/268 on farming in less favoured agricultural areas.

15   House of Lords Select Committee on the European Communities, *Agriculture and the Environment*, 20th Report Session 1983–1984 (London: HMSO).

16   Haigh, N. (1987), *EEC Environmental Policy and Britain*, 2nd ed. (Harlow: Longman Group UK Ltd) 311.

17   *Ibid.* The quotation is taken from the announcement made by Lord Belstead, then Minister of Agriculture, as reported by Haigh.

agricultural structures.[18] Article 19 of this Regulation made provision for Member States to introduce agri-environment measures, such as the United Kingdom's Broads Grazing Marshes Conservation Scheme.

Article 19 gave Member States discretion to introduce agri-environment schemes within nationally designated 'environmentally sensitive areas'. These were defined as being areas of 'recognised importance from an ecological and landscape point of view.' Within such areas Member States could make financial payments to farmers who undertook to farm in a manner that preserved or improved the environment. Article 19 required that this undertaking should oblige farmers to avoid agricultural intensification and to ensure that both their stock densities and the level of intensity of their general farming practice were compatible with the specific environmental needs of their land.

Regulation 797/85 made no provision for European Community funding to be provided to support measures adopted by Member States adopted under Article 19. This was a major disincentive and only four Member States, initially, introduced agri-environment schemes.[19] Subsequently, Regulation 1760/87 authorised the Community to reimburse 25 per cent of Member States' eligible expenditure.[20] Even so, the scheme continued to operate in only four Member States.[21] However, this limited application came to an end in 1992, when Regulation 2078/92 made it compulsory for all Member States to implement agri-environment measures.[22]

Transforming Agri-Environment Policy into a Central Plank of Rural Development Policy

In making agri-environment measures compulsory, Regulation 2078/92 marked an important development in the role of the scheme. This Regulation was, subsequently, repealed and replaced by Regulation 1257/99,[23] which itself has been repealed and replaced by Regulation 1698/2005.[24] These later Regulations developed the Community's agricultural structures policy into a broader rural development

---

18   Council Regulation 797/85, [1985] OJ L95/1.

19   See Baldock D., Lowe, P., footnote 9 above, 16. The specific Member States were Denmark, Germany, the Netherlands and the United Kingdom.

20   Council Regulation 1760/87 [1987] OJ L167/1.

21   European Commission, *Proposal for a Council Regulation on the Introduction and Maintenance of Agricultural Production Methods Compatible with the Requirements of the Protection of the Environment and the Maintenance of the Countryside*, COM (1990) 366 Final. Once again the four Member States concerned were Denmark, Germany, the Netherlands and the United Kingdom.

22   Council Regulation 2078/92 [1992] OJ L215/1.

23   Council Regulation 1257/99 on support for rural development, [1999] OJ L160/80.

24   Council Regulation 1698/2005 on support for rural development, [2005] OJ L277/1.

policy. They also required Member States, or their regional authorities, to design multi-annual rural development plans, most recently for the period 2007 to 2013. It also continued to be compulsory for Member States to include agri-environment measures within each rural development plan. Currently, these measures must require participating farmers to enter agri-environmental commitments for between five and seven years, though longer-term agreements can also be concluded where this can be shown to be 'necessary and justified'.[25]

Given the broad diversity of agriculture industry across the Community and the different landscape and climatic conditions in which agriculture is practised, it is, perhaps, not surprising that Member States have been given broad discretion in constructing their agri-environmental measures. In 1992, Regulation 2078/92 explained that the Community's objective was to promote:[26]

a. The use of farming practices which reduce the polluting effects of agriculture...
b. An environmentally favourable extensification of crop farming, and sheep and cattle farming, including the conversion of arable land into extensive grassland
c. Ways of using agricultural land which are compatible with protection and improvement of the environment, the countryside, the landscape, natural resources, soil and genetic diversity
d. The upkeep of abandoned farmland and woodlands where this in necessary for environmental reasons or because of natural hazards and fire risks and thereby avert the dangers associated with the depopulation of agricultural areas
e. Long term set aside of agricultural land for reasons connected with the environment
f. Land management for public access and leisure activities
g. Education and training for farmers in types of farming compatible with requirements of environmental protection and upkeep of the countryside

More recently, Regulation 1698/2005 simply provides that 'agri-environment payments shall be granted to farmers who make, on a voluntary basis, agri-environmental commitments.'[27] The only prescription is that these commitments must go beyond the mandatory environmental obligations already introduced by the cross-compliance measures contained in Regulation 1782/2003.[28]

---

25   *Ibid.* Article 39(4).

26   Council Regulation 2078/92, Article 1.

27   Council Regulation 1698/2005 Article 39(2).

28   See Chapter 3 for an examination of these cross-compliance measures. See also the Commission implementing regulation, Commission Regulation 1974/2006 [2006] OJ L368/1.

As a consequence of the level of discretion conferred upon Member States, agri-environment schemes have varied widely from Member State to Member State and, indeed, in some cases from region to region within one Member State. For example, Table 5.1 shows the European Commission's analysis of Community expenditure on the measures adopted by Member States under Regulation 2078/92.[29] For the purposes of the analysis, the European Commission sub-divided Member States' measures into 5 categories:

1. Environmentally beneficial productive farming
   a. Organic farming;
   b. Non organic farming with environmental improvements;
   c. Maintenance of existing low intensity systems.
1. Non Productive land management (20 year set aside, maintenance of abandoned land, landscape features, public access etc.).
2. Training and Demonstration projects

Table 5.1   Estimated Proportion of Budgeted Spending in Each Member State as percentage of Category of Measure (1996 Programme)

Measure Type/ funding %	B	Dk	D	El	E	F	Irl	I	NL	L	Ös	P	Fin	S	UK
1(a)	20	24	1	14	4	3	2	23	2	1	17	4	5	15	2
1(b)	58	46	56	35	35	15	49	43	32	39	59	18	42	6	53
1(c)	5	16	21	0	15	79	21	22	0	56	21	68	42	71	30
2	14	14	21	50	42	3	24	10	0	3	3	6	7	1	14
3	3	0	1	0	4	1	4	2	66	0	0	4	5	7	0

Source: European Commission, *Report on the Application of Council Regulation 2078/92*, COM (1997) 620 © European Communities 1997. Only European Community legislation printed in the paper version of the Official Journal of the European Union is deemed authentic.

This table illustrates the different approaches adopted by Member States. For example, the Netherlands chose to focus its scheme upon training and

    29   European Commission, *Report on the Application of Council Regulation 2078/92*, COM (1997) 620.

demonstration projects.[30] In contrast, such training measures formed no part of the schemes then implemented in Denmark, Greece, Luxembourg, Austria and the United Kingdom. Similarly, whilst Finland, France, Luxembourg, Portugal, Sweden and some Länder in Germany introduced substantial measures to protect existing extensive farming practices, no such measures were adopted within Greek or Dutch agri-environment schemes.[31]

Buller has also pointed to the emergence of differing agri-environmental policy agendas.[32] In particular, he identifies a north/south divide, explained by differing levels of agricultural modernisation and intensification in northern and southern Member States. While agri-environmental measures in northern European Community Member States predominantly focused upon nature and landscape protection and the prevention of farm-based pollution, those in southern European Member States often emphasised the need to support marginal agricultural activities and thus avoid land abandonment and the loss of extensive farmland to natural recolonisation or afforestation.[33]

Variations have also developed in the territorial scope of Member States' agri-environment schemes. When the scheme was introduced in 1985, Member States were only able to apply it within nationally designated 'environmentally sensitive areas' of recognised ecological or landscape importance.[34] Subsequently, Regulation 2078/92 made it compulsory for Member States to implement the scheme throughout their territories.[35] In doing so, however, Member States were given a choice. They could divide their territories into separate zones, introducing different measures in each zone, or they could introduce one or more agri-environmental measure throughout their national territory.[36] More recently both Regulation 1257/99[37] and Regulation 1698/2005[38] have stipulated that Member States should provide agri-environment schemes throughout their territories 'in accordance with their specific needs'. This enabled Member States to continue choosing between uniform measures applying throughout their territory and zonal measures applying in particular areas. One outcome of this approach has been the huge number of agri-environmental schemes introduced by Member States. The European Commission originally anticipated that one agri-environment scheme

---

30   *Ibid.* 9.

31   *Ibid.*

32   Buller, H. (2000), Regulation 2078: Patterns of Implementation, in Buller, H., Wilson, G.A., Höll, A. (2000), *Agri-environmental Policy in the European Union* (Aldershot: Ashgate Publishing Ltd).

33   *Ibid.* 224.

34   Council Regulation 797/85, Article 19.

35   Council Regulation 2078/92, Article 3(1).

36   *Ibid.* Article 3(1)–3(3). Article 3(2) provides that each zonal programme should cover an area that is homogenous in terms of the environment and the countryside.

37   Council Regulation 1257/99, Article 43(2).

38   Council Regulation 1698/2005, Article 39(1).

would be introduced in each Member State.[39] However, in reality, it had approved some 158 individual schemes by late 1998.[40] One distinction between these schemes, highlighted by the European Commission, has been between 'broad and narrow' schemes.[41] The Commission noted that broad measures usually attract a large number of farmers, cover a wide area, make modest demands on farmers and pay relatively little for the environmental services provided by these farmers.[42] In contrast, narrow schemes usually target site-specific environmental issues, have fewer farmers enrolled upon them, make substantial demands on farming practices and make higher payments for the environmental services provided.[43] As Potter has observed, some 'broad' schemes have been open to the criticism that they have enabled Member States to make payments to a large number of farmers, without any consideration for the environmental value of their farms or for the environmental benefits that their participation in the scheme has achieved.[44] In particular, in this regard, he identified France's *prime à l'herbe* scheme and Austria's grassland extensification scheme.[45] Both of these schemes applied throughout these Member States and sought to encourage livestock extensification. However, they were equally open to farmers who might have had no intention of intensifying their farming.

Broad differences have also emerged in the levels of payments made to farmers participating in agri-environmental schemes. In 2001, average annual payments varied from €246 per hectare in Greece to just €32 per hectare in France.[46] To some extent, some differences are to be expected since individual schemes make very different demands of farmers and, consequently, payments will vary from scheme to scheme. Payment levels are fixed by Member States, subject to ceilings

39 See Evidence of Rheinhard Priebe, Head of the Agri-Environment Programme – European Commission, to the House of Commons Select Committee, *Environmentally Sensitive Areas and Other Schemes Under the Agri-Environment Regulation*, 2nd Report, Session 1996–1997, HC45-11, 273.

40 Court of Auditors, Special Report 14/2000, *Greening the CAP*, OJ [2000] C353/1, 18.

41 European Commission (2005), *Agri-Environment Measures: Overview on General Principles, Types of Measures and Application*, Brussels: European Commission, 10. See also Buller, H., footnote 32 above, 233.

42 *Ibid.*

43 *Ibid.*

44 Potter, C., (1998), *Against the Grain: Agri-Environmental Reform in the United States and the European Union*, Wallingford: CAB International, 125.

45 *Ibid.*

46 European Commission (2004), *The Agricultural Situation in the European Union: 2003 Report* (Luxembourg: Office for Official Publications of the European Communities), table 3.6.2.3. The average payments for Ireland and the United Kingdom were €131 and €104 per hectare respectively.

set out within the Regulation.[47] The European Community also stipulates the criteria upon which the payments are calculated. Regulation 1257/99 provided for payment levels to be calculated on the basis of farmers' income foregone as a result of participating in the scheme, together with additional costs resulting from commitments given under the scheme and also the need to provide a financial incentive to entice farmers into the scheme.[48] Commission Regulation 445/2002 made it clear that the value of any incentive element could generally not exceed 20 per cent of the income foregone and of the additional costs incurred by participating farmers.[49] More recently, in an attempt to ensure that the scheme complied with the provisions of the WTO Agriculture Agreement,[50] the scope to make these incentive payments has been removed. Under Regulation 1698/2005 payments are to be calculated solely on the basis of the income foregone by farmers as a result of taking part in the scheme and also any additional costs arising from their participation.[51] This raises the concern that focusing solely upon income foregone creates a northern European bias. Buller has pointed out that the principal objective of southern European Member State agri-environment schemes has been to prevent land abandonment.[52] But, if payment levels are to be determined principally by income foregone as a result of joining an agri-environmental scheme then this is unlikely to assist in achieving that goal. Too often, low farm income was the catalyst that led farmers to decide to leave the land.

One final distinction concerns the financial contribution made to Member States by the European Community to help defray the cost of these measures. Although the Community initially provided no financial assistance at all, Regulation 1760/87 provided for 25 per cent of Member States' eligible expenditure upon agri-environmental measures to be reimbursed. Subsequently, under Regulation 2078/92, the European Community agreed to make greater sums available to Member States. Under that Regulation it would generally reimburse 50 per cent of Member State spending and cover 75 per cent of the costs in convergence

---

47   Council Regulation 1698/2005, Article 39(4) and Annex sets maximum annual payment levels in relation to particular agricultural land uses: annual crops: €600 per hectare, specialised perennial crops: €900 per hectare, other land used: €450 per hectare, local breeds in danger of extinction: €250 per hectare.

48   Council Regulation 1257/99, Article 24.

49   Commission Regulation 445/2002 laying down detailed rules for the application of Council Regulation 1257/99 on support for rural development, [2002] OJ L74/1. Exceptionally, under Article 19, the 20 per cent cap on incentives could be breached if a higher incentive was deemed to be indispensable to ensure the effective implementation of specific commitments.

50   See further Chapter 8.

51   Council Regulation 1698/2005, Article 39(4). See also Commission Regulation 1974/2006 laying down detailed rules for the application of Council Regulation 1698/2006 ([2006] OJ L368/1), Article 53.

52   Buller, H., footnote 32 above, 224.

areas (formerly known as objective one regions).[53] More recently, the Community increased, again, the percentage of national expenditure that it will refund. Today, it will refund up to 60 per cent of eligible national spending, rising to up to 85 per cent in convergence areas.[54]

The Environmental Credentials Agri-Environment Policy

The European Community, clearly, regards its agri-environment scheme as being a success. In 1993, its 5th Environmental Action Plan set a target of enrolling 15 per cent of the European Community's agricultural area into the scheme by 2000.[55] In reality, some 20 per cent of this land was enrolled by 1999.[56] However, this alone is not an accurate measure of the scheme's success. In particular, this figure hides the wide disparities that exist in levels of enrolment in different Member States. By 2002, virtually all farmland in Luxembourg and Finland had been enrolled within the scheme, in contrast to less than 10 per cent of farmland in Italy and Spain and less than 5 per cent of farmland in Greece and the Netherlands.[57] The European Commission has identified a number reasons for these variations – the degree of experience with agri-environment measures in each Member State, the attitude to these schemes at all levels, the knowledge base available in relation to agri-environmental farming, the budget available and the level of payments available for farmers taking part in the scheme.[58]

The attitude of national policy makers has been crucialin determining the importance to be attached to the scheme in individual rural development plans. This is another area where broad differences exist between Member States. Agri-environmental measures account for almost 90 per cent of the European Community's rural development expenditure in Sweden, but less than 10 per cent of such expenditure in Greece.[59] The willingness of national treasuries to provide funding to match that allocated by the European Community also continues

---

53   Under Council Regulation 2052/88 ([1988] OJ L185/9) (subsequently repealed and replaced by Council Regulation 1260/99 ([1999] OJ L161/1) objective one areas were areas whose gross domestic product was less than 75 per cent of the European Community average.

54   Council Regulation 1783/2003 amending Council Regulation 1257/99 on support for rural development. [2003] OJ L270/70, article 1.

55   See Council Resolution on a Community action programme of policy and action in relation to the environment and sustainable development, [1993] OJ C138/38.

56   European Commission (1999), *Agriculture, Environment, Rural Development: Facts and Figures – A Challenge for Agriculture* (Luxembourg: Office for Official Publications of the European Communities) 121.

57   European Commission (2005), *Agri-Environment Measures: Overview on General Principles, Types of Measures and Application* (Brussels: European Commission) 7.

58   *Ibid.* 9.

59   *Ibid.* 6.

to play an important role. It also has an important influence upon the levels of payment available. In this regard, the scheme continues to be in direct competition with agricultural production policy. Farmers will only enrol if the payment levels available under the scheme in their Member State adequately compensate them for any production-based payments they may lose as a result. It does not help that rural development policy, in general, continues to be the poor cousin within the Common Agricultural Policy. Of the total Community expenditure on agriculture 22 per cent is used for rural development,[60] with 44 per cent of this being allocated to agri-environment schemes.[61] The rest of the Community's agricultural expenditure is used, principally, to fund the direct payments made within agricultural production policy. The 22 per cent figure represents a welcome increase on the 5 per cent of the Community's agricultural spending allocated to agriculture's structural policy in the past.[62] However, Community agri-environment spending continues to be dwarfed by spending upon production policy. This disparity is accentuated by the fact that production policy measures are fully funded by the Community, whilst rural development measures are only part-funded by it.

The agri-environment scheme forms an important element of the European Community's strategy to integrate environmental protection requirements into the CAP.[63] It also plays an important part in supporting wider European Community environmental policy commitments. The 2008 CAP Health Check pointed to its role guarding against climate change, in helping the Community to achieve its goal of halting biodiversity loss by 2010 and in delivering the water quality and quantity targets set by the Water Framework Directive.[64] In reality, however, the scheme can, presently, only be considered to be making a partial contribution towards such goals. Indeed, in Member States with low rates of enrolment in the scheme, agri-environmental measures make very little contribution to the integration of environmental protection issues within agricultural practice or to the protection of the wider environment.

In terms of the integration of environmental protection requirements into the CAP, the 'polluter pays' principle is one of the central strands of the European Community's environmental policy. However, the European Commission has

---

60   European Commission (2008), *The Agricultural Situation in the European Union, 2007 Report* (Luxembourg: Office for Official Publications of the European Communities) table 3.4.3.1.

61   European Commission (2008), *Rural Development in the European Union: Statistical and Economic Report 2007* (Luxembourg: Office for Official Publications of the European Communities) 19.

62   Snyder, F.G. (1985), *Law of the Common Agricultural Policy* (London: Sweet and Maxwell Ltd) 164.

63   As required by Article 6 of the EC Treaty.

64   In relation to the European Community's goal of halting biodiversity loss by 2010 see the Council Decision adopting the Community's 6th Environmental Action Plan, [2002] OJ L242/1, Article 6(1). For further discussion on biodiversity protection and the prevention of water pollution see later, Chapters 7 and 8.

estimated that, in 2002, some 26 per cent of land entered in agri-environment agreements across the Community had been enrolled in national agreements that principally provided payments to farmers in return for their commitment to reduce farm inputs, such as fertilisers and pesticides.[65] This raises questions about the compatibility of these measures with the 'polluter pays' principle.[66] Also, it does not equate with the European Commission's philosophy that farmers should be expected to observe basic environmental standards without compensation and agri-environmental measures should be used to purchase additional environmental services.[67] The introduction of compulsory cross-compliance measures, in 2003, helped to move the CAP more closely towards this philosophy. Regulation 1698/2005 now provides that agri-environmental payments will only cover commitments going beyond the mandatory cross-compliance standards imposed by Regulation 1782/2003.[68] In principal, European agriculture has embarked upon a twin-track approach. In areas of high quality land farmers are likely to farm intensively in order to maximise their incomes from production. Few important environmental features may remain on these lands. However, these farmers are now be required to avoid causing pollution and environmental damage as part of their cross-compliance commitments. In contrast, areas of lower quality agricultural land will, usually, have experienced less intensification and more environmental features are likely to remain. The lower production potential of these areas also means that farmers there can be expected to be more attracted to an agri-environment scheme. In practice, the separation between commitments forming part of national cross-compliance standards, particularly the nationally determined good agricultural and environmental conditions, and those for which payment is available under agri-environment schemes is not always clear.[69] This creates the potential that farmers in one Member State may receive payments for providing services that are expected of them and, but other may go unrewarded in another Member State.

Question exist over the contribution that the scheme has made, so far, toward the European Community's broader environmental goals. In relation to the protection of biodiversity, a European Commission assessment of agri-environmental measures in place, in 2002, suggested that only 15 per cent of land enrolled was

---

65   European Commission, footnote 57 above, 13.

66   See Court of Auditors, Special Report 14/2000: *Greening the CAP*, OJ [2000] C353/1, 36 and Dhont, N. (2003), *Integration of Environmental Protection into Other EC Policies* (Groningen: Europa Press) 272.

67   European Commission, footnote 1 above.

68   Council Regulation 1698/2005, Article 39(3). A similar provision could also be found in Article 23(2) of Council Regulation 1257/99, though, as noted in Chapter 3, compulsory cross-compliance did not apply at that time.

69   See Court of Auditors, Special Report 8/2008, *Is Cross Compliance an Effective Policy?* Available at   http://eca.europa.eu/portal/page/portal/publications/ auditreportsandopinions/specialreports. Search under 2008 to access the report.

taking part in measures that specifically targeted biodiversity and landscape enhancement.[70] This relatively low emphasis upon biodiversity and landscape enhancement suggests that national agri-environment measures have not always promoted the management practices that would be most effective in addressing the Community's broader environmental objectives, such as halting biodiversity loss. This is also borne out by an examination of areas of high nature conservation value. Several reports have highlighted the fact that areas which traditionally supported low intensity agriculture tended to have high nature conservation value farming systems.[71] Large areas of such farmland are found in Spain and Portugal. However, the European Commission has reported that less than 10 per cent of land enrolled in an agri-environment scheme in these countries is participating in schemes specifically targeting nature and landscape enhancement.[72] This is in contrast with a figure of 50 per cent in the Netherlands, a Member State perhaps better known for its intensive farming. Clearly, Member States need to develop sufficiently targeted agri-environment measures to enable the Community, also, to achieve its own policy goals for the scheme. The 2008 Health Check Reforms may represent a first step in that direction. From 1 January 2010, Member States must ensure that their rural development programmes provide, in accordance with their own needs, for operations that will support the European Community's environmental priorities, namely climate change, renewable energies, water management, biodiversity, innovation in these policy areas and also restructuring within the dairy sector.[73] The Community has also adopted an indicative list of potential rural development measures that can be utilised to help address these issues.[74] This makes it clear that agri-environment agreements have an important contribution to make in helping the Community to achieve its policy goals on climate change, water management, the protection of biodiversity and dairy restructuring. Elsewhere the Community's Strategic Guidelines for Rural Development provide:[75]

---

70   European Commission, footnote 57 above, 13.

71   See Baldock, D. et al. (1993), *Nature Conservation and New Directions in the EC Common Agricultural Policy* (London: Institute for European Environmental Policy); Beaufoy, G. et al. (1994), *The Nature of Farming: Low Intensity Farming Systems in Nine European Countries* (London: Institute for European Environmental Policy) or Bignal, E., McCracken, D., Ecological Resources of European Farmland, in Whitby, M. (ed.) (1996), *The European Environment and CAP Reform, Policies and Prospects for Conservation* (Wallingford: CAB International).

72   European Commission (2004), Working Document: *Biodiversity Action Plan for Agriculture – Implementation Report* (Brussels: European Commission Agriculture Directorate General) 10.

73   Council Regulation 1698/2005, [2005] OJ L277/1, Article 16(a) as inserted by Council Regulation 74/2009, [2009] OJ L30/100.

74   *Ibid.* Annex II.

75   See Council Decision 2006/144 on the Community's Strategic Guidelines for Rural Development (Programme 2007–2013), [2006] OJ L55/20, as amended by Council Decision 2009/61, [2009] OJ L30/112.

Under Axis 2, the agri-environment measures and forestry measures can be used in particular to enhance biodiversity by conserving species rich vegetation types and protecting and maintaining grassland and extensive forms of agricultural production. Specific actions under Axis 2, such as agri-environment measures or afforestation, can also help to improve the capacity to better manage the available water resources in terms of quantity and protect them in terms of quality. Furthermore, certain agri-environmental and forestry actions contribute to curbing emissions of nitrous oxide and methane and help to promote carbon sequestration.

The Community's position is, therefore, very clear. For their part, Member States must adopt revised rural development plans reflecting these issues by 30 June 2009.

However, the ball remains firmly in the Member States' court when it comes to turning these objectives into practical reality.

An Evaluation of Agri-Environmental Measures in the United Kingdom

Following the introduction of Regulation 797/85, section 18 of the Agriculture Act 1986 provided for the Minister for Agriculture to designate Environmentally Sensitive Areas (ESAs), in the United Kingdom, where this would facilitate the conservation and enhancement of the natural beauty of the area, the conservation of the flora, fauna, geological/physiological features of the area or protection of buildings or other objects of archaeological, architectural or historical interest.[76] Within two years, 19 ESAs had been designated across the United Kingdom: 10 in England, 5 in Scotland, 2 in Wales and 2 in Northern Ireland.[77] Farmers located within ESA boundaries had the opportunity to enter all or part of their farms into voluntary 10 year management agreements with MAFF, with a break clause allowing either to terminate after 5 years. The obligations imposed by these agreements were tiered. All participating farmers agreed to observe basic requirements imposed at tier one. Essentially, they agreed not to intensify their farming, by not increasing livestock numbers or fertiliser and pesticide use, and they agreed not to conduct farm developments such as land reclamation, draining or ploughing and reseeding unimproved or uncultivated land. Most ESAs offered farmers the opportunity to enter the scheme at a higher tier, where higher payments

---

76   In Wales this power was exercised by the Secretary of State for Wales, whilst the Agriculture (Environmental Areas) (Northern Ireland) Order 1987, SR 1987/458, made corresponding provision for Northern Ireland.

77   Hart, K., Wilson, G.A. (2000), United Kingdom: from agri-environmental policy shaper to policy receiver, in Buller, H., Wilson, G.A. (eds) (2000), *Agri-Environmental Policy in the European Union* (Aldershot: Ashgate Publishing Ltd) 103.

reflected more stringent management conditions. Payments were also available for carrying out 'capital activities' such as restoring hedges or stonewalls.

The initial ESA scheme fitted very well within the United Kingdom's traditional approach to nature conservation, providing targeted protection to habitats within small areas.[78] But Regulation 2078/92 required a new approach, which addressed the need to deliver environmental protection within the wider countryside. The United Kingdom responded by expanding its ESA scheme and by introducing several new schemes. The number of ESAs rose from 19 in 1988 to 43 in 1997, of which 22 were in England, 10 in Scotland, 6 in Wales and 5 in Northern Ireland.[79] A structural weakness within the initial ESA scheme was also addressed. This was the fact that farmers, initially, could enrol a part of their farm within the scheme. This created a 'halo effect', where some farmers received ESA payments for avoiding intensification upon part of their farms, whilst, simultaneously, intensifying production on farmland not enrolled within the scheme.[80] To counter this problem it became compulsory for farmers to enrol their entire farms within the scheme.

Outside designated ESAs, a Countryside Stewardship scheme was introduced in England in 1991,[81] whilst Scotland and Northern Ireland introduced Countryside Premium and Countryside Management Schemes in 1997 and 1999 respectively.[82] Unlike the ESA scheme, which was open to every farmer within each ESA area, eligibility for these schemes was determined by a tendering process. Applicants were ranked by their farm's environmental potential and only those with the greatest potential benefit accepted. Successful applicants, again, entered 10 year management agreements with MAFF, with a break clause at 5 years. On this occasion, however, payments were made in return for providing conservation management of the particular habitats or landscape features found on their farms, with further payments, again, available for capital activities.

In addition, a number of other pilot schemes were also introduced. Outside ESA areas, a range of habitat schemes encouraged long term set aside of farm land and its management as wildlife habitat. In England, this operated on land adjacent

---

78   *Ibid.* 101.

79   House of Commons Agriculture Committee, 2nd Report, Session 1996–1997, *Environmentally Sensitive Areas and Other Schemes Under the Agri-Environment Regulation*, HC-45-I, xviii.

80   Hart K., Wilson, G.A., footnote 77, 105.

81   Introduced as a pilot scheme by the Countryside Commission, responsibility for it was subsequently transferred to MAFF under the Countryside Stewardship Regulations 1996, SI 1996/695.

82   See the Countryside Premium Scheme (Scotland) Regulations 1997, SI 1997/330 (S.27) and the Countryside Management Regulations (Northern Ireland) 1999, SR 1999/208. Wales introduced a pilot Tir Cymen scheme which operated in only three areas: Meironydd, Dinefwr and Swansea. It made payments in return for farmers' compliance with a code of practice, with additional payments being available for recognised capital projects.

to particular waterways,[83] farmland that had participated in the 1988 voluntary set aside scheme[84] and farmland located in areas of coastal salt marsh. Separate habitat schemes in Scotland,[85] Wales[86] and Northern Ireland[87] targeted other habitats. Also, outside ESA areas, a moorland extensification scheme provided payments to farmers who observed winter and summer stocking densities on moorland areas. [88] Additionally, a Countryside Access scheme provided payments to farmers who agreed to manage farmland to promote public access for walking and recreation. [89] Finally, an organic farming scheme sought to encourage farmers to convert to organic farming. [90] In this case the scheme provided annual payments over a five year period, based upon the area of the farm, tapered to gradually reduce over this conversion period.

The United Kingdom played a central role in the initial development of European Community agri-environment policy. However, it has been criticised for having 'merely responded to Regulation 2078 as a 'receiver' of Brussels policy, rather than acting as a 'shaper' of the Regulation as was the case earlier for Regulation 797/85.'[91] Although a range of new measures was introduced, the ESA scheme remained the flagship of United Kingdom agri-environmental policy. For example, 61 per cent of agri-environmental spending in England in 1997 went on

---

83   The Habitat (Water Fringe) Regulations 1994, SI 1994/1291, applied to farmers with land adjacent to the Derwent and Rye Rivers in Yorkshire, the River Medway in Kent, the Slapton Lay in Devon and tributaries of the River Ribble in Lancashire.

84   The Habitat (Former Set Aside) Regulations 1994, SI 1994/1292.

85   The Habitats (Scotland) Regulations 1994, SI 1994/2710 (S.138), targeted waterside habitats, upland scrub, coastal heaths, damp lowland grassland or marsh and dry lowland grassland.

86   The scheme targeted species rich grasslands, water fringe habitats, coastal lands and broad leaved woodland. See the Habitats (Broadleaved Woodland) (Wales) Regulations 1994, SI 1994/3099; the Habitats (Coastal Belt) (Wales) Regulations 1994, SI 1994/3101; the Habitats (Species Rich Grassland) (Wales) 1994, SI 1994/3102 and the Habitats (Water Fringe) (Wales) Regulations 1994, SI 1994/3100.

87   The Habitat Improvement Regulations (Northern Ireland) 1995 targeted designated water fringes, farmland adjoining Areas of Special Scientific Interest and farmland linking together important wildlife habitats and grazed semi-natural woodland.

88   By the Moorland (Livestock Extensification) Regulations 1995, SI 1995/904; the Moorland (Livestock Extensification) (Wales) Regulations 1995, SI 1995/1159; the Heather Moorland (Livestock Extensification) (Scotland) Regulations 1995/891 (Scotland; 74), the Moorland (Livestock Extensification) (Northern Ireland) Regulations 1995, SR 1995/239.

89   For England and Wales see the Countryside Access Regulations 1994, SI 1994/2349, elsewhere see the Set Aside Access (Scotland) Regulations 1994, SI 1994/3085 (S.165) and the Countryside Access Regulations (Northern Ireland) 1996, SR 1996/23.

90   Under the Organic Farming (Aid) Regulations 1994, SI 1994/1721; The Organic Aid (Scotland) Regulations 1994, SI 1994/1701 (S.75) and the Organic Farming Aid Regulations (Northern Ireland) 1999, SR 1999/663.

91   K. Hart and G.A. Wilson, footnote 77 above, 116.

the ESA scheme, 26 per cent on the Countryside Stewardship and 13 per cent on the remaining agri-environmental measures.[92]

An evident problem was the limited matched funding available from the Treasury. The United Kingdom allocated fewer funds to agri-environmental policy than many other Member States.[93] This, inevitably, limited the breadth of the measures implemented and the number of farmers participating. Certainly, the European Commission observed that, in 1999, 11 per cent of farmers in the United Kingdom were taking part in agri-environmental measures, which compared poorly with participation rates of 23 per cent in France and 78 per cent in Austria.[94]

The vast majority of farmers participating in agri-environmental measures in the United Kingdom had enrolled in the ESA scheme. In England, 87 per cent of eligible land within ESA areas had enrolled within the scheme within its first year.[95] However, the vast majority of these farmers did so only at tier one. Approximately 85 per cent of farmers who entered the scheme did so at this level, effectively agreeing not to intensify their production levels and to retain any landscape and habitat features then existing upon their farms.[96] Few, therefore, made the commitments to manage and enhance habitat features, as would have been required through joining at a higher tier. Consequently, English Nature observed, 'ESAs are particularly well adapted to secure and sustain the broader fabric of countryside character over large areas. They have perhaps been less successful in the enhancement and re-creation of the biodiversity resource.'[97] For many farmers, participation in the scheme did not, actually, require any significant alterations to their farming practice. To some extent this reflected the mixed objectives of Regulation 2078/92. The scheme's stated objectives suggested that it was as much about protecting farm incomes as about environmental protection.[98] It might be argued that, through enrolling, these farmers could, in future, develop more

---

92   House of Commons Agriculture Select Committee, footnote 79 above, xxiv.

93   This is evident from the levels of Community spending in individual Member States, which was only available when Member States made supporting payments. In 1996 the European Community paid 232 million ecu to co-finance agri-environmental measures in Germany, 119 million ecu to do so in France, but only 25.5 million ecu to co-finance United Kingdom agri-environmental measures. See Hart K., Wilson G.A., footnote 77 above, 116.

94   European Commission (1999), *Agriculture, environment, rural development: Facts and Figures: A Challenge for Agriculture* (Luxembourg: Office for Official Publications of the European Communities) 124.

95   Whitby, M., Lowe, P. The Political and Economic Roots of Environmental Policy in Agriculture, in Whitby, M. (1994), *Incentives for Countryside Management: The Case of Environmentally Sensitive Areas* (Wallingford: CAB International) 18.

96   Ministry of Agriculture, Fisheries and Food, Evidence to the House of Commons Select Committee on Agriculture, footnote 79 above.

97   English Nature, Evidence to the House of Commons Select Committee on Agriculture, footnote 79 above, 124.

98   See Council Regulation 2078/92, article 1.

positive attitudes towards conservation and become willing to accept greater land management obligations. However, research suggests that most farmers simply joined the scheme to qualify for payments, with little evidence that they were subsequently motivated to enter higher ESA tiers. [99]

The Countryside Management Scheme also proved popular with farmers. The scheme's tendering system ensured that the monies available were used to enrol the most environmentally valuable applications. However, the scheme's limited budget also meant that many good quality applications were refused. This risked a loss of goodwill and, generally, showed a lack of genuine commitment towards an agri-environmental policy in the wider countryside. [100]

Beyond the Environmentally Sensitive Area scheme and the Countryside Management scheme, the other agri-environment schemes were failures. At one level farmers proved unwilling to accept long term set aside under the Habitat options. [101] Additionally, the simple truth was that these schemes, like all agri-environment schemes, were in direct competition with the European Community's agricultural production policies. Farmers would only enrol in these schemes if they at least matched any production payments they would lose through doing so. Too often, the United Kingdom schemes did not do so. [102] Instead of being integrated into agricultural production policy, agri-environmental measures were in direct competition with it. For example, the House of Commons Agriculture Committee reported that the Moorland Livestock Extensification scheme 'foundered on the simple economics of upland sheep rearing.' [103] Similarly, in relation to negligible interest in conversion to organic farming, it noted that the levels of payment for conversion in the United Kingdom were the lowest in the European Community, whilst it also pointed to the Countryside Access scheme as being a 'dismal failure' almost completely shunned by farmers. [104]

The agri-environment measures available to farmers today in the United Kingdom are very different from those introduced under Regulation 2078/92. The Habitat, Moorland Livestock Extensification and Countryside Access schemes failed to attract sufficient numbers of applications from farmers and were withdrawn in the late 1990s. Each United Kingdom region continues to encourage farmers to convert to organic methods. In England, payments to organic farmers both during

---

99   Morris, C., Potter, C. (1995), Recruiting the new conservationists: farmers' adoption of agri-environmental schemes in the United Kingdom. [1995] 11 *Journal of Rural Studies* 51–63. See also Potter, C. (1998), *Against the Grain, Agri-Environmental Reform in the United States and European Union* (Wallingford: CAB International) 89.

100   See K. Hart and G.A. Wilson, footnote 77 above, 110.

101   House of Commons Agriculture Committee, footnote 79 above, xxxi, see also page xxxii for similar complaints about the inflexibility of the Moorland Livestock Extensification Scheme.

102   *Ibid.* xxxvi.

103   *Ibid.* xxxi.

104   *Ibid.* xxxii and xxxiii respectively.

and after their conversion period, form part of the broader agri-environmental programme.[105] In Scotland and Wales these payments are made through separate organic farming schemes.[106]

In designing agri-environment measures for their 2007–2013 Rural Development Plans, each agriculture authority removed distinctions between Environmentally Sensitive Areas and other areas. Instead, their agri-environmental schemes now apply throughout each UK region. These are as follows:

England:	The Environmental Stewardship Scheme[107]
Scotland:	The Rural Stewardship Scheme[108]
Wales:	Tir Gofal and Tir Cynnal[109]
Northern Ireland:	The Countryside Management Scheme[110]

Each scheme requires farmers to comply with, either the general cross-compliance conditions introduced within CAP production policy,[111] or with specific environmental conditions introduced as part of the agri-environment scheme.[112] Additionally, participating farmers must now accept responsibility for actively managing particular habitats or features. These management requirements address the criticisms made of earlier schemes; that farmers could qualify for payments by simply agreeing not to intensify production but without having to actively manage the farmland environment. For example, the Environmental Stewardship scheme provides a menu of 66 management options across a range of farm habitats and

---

105   Known as Organic Entry Level Stewardship.

106   In Northern Ireland, farmers who achieve organic status can receive payments as part of an agri-environment scheme, the Countryside Management scheme, whilst those undergoing conversion are eligible for payments through a separate organic farming scheme.

107   See the Environmental Stewardship (England) Regulations 2005, SI 2005/621, as amended by the Environmental Stewardship (England) and Countryside Stewardship (Amendment) Regulations 2006, SI 2006/991 and the Environmental Stewardship (England) and Organic Products (Amendment) Regulations 2006, SI 2006/2075.

108   See the Rural Stewardship (Scotland) Regulations 2001, SI 2001/300 as amended by the Rural Stewardship (Scotland) (Amendment) Regulations 2003, SI 2003/300, the Rural Stewardship (Scotland) (Amendment No.2) Regulations 2003, SI 2003/1717 and the Rural Stwardship (Scoatland) (Amendment) Regulations 2005, SI 2005/620.

109   See the Land in Care (Tir Gofal) (Wales) Regulations 1999, SI 1999/1176 as amended by the Tir Gofal (Wales) (Amendment) Regulations 2006, SI 2006/1717 and the Tir Cynnal (Wales) Regulations 2006, SI 2006/41.

110   See the Countryside Management Regulations (Northern Ireland) 2008, SR 2008/174.

111   As in the case of Environmental Stewardship in England and Countryside Management in Northern Ireland.

112   As with Rural Stewardship in Scotland and both Tir Gofal and Tir Cynnal in Wales.

features, with individual points being ascribed to each option.[113] Farmers wishing to enter the scheme must then reach a target of 30 points per hectare for their farms, or 15 points per hectare if the agricultural land is situated within a less favoured area.[114] Alternatively, farmers can also be enrolled in order to undertake a 'special project', which enables farmers to enter the scheme in return for undertaking a management project not otherwise provided for by the scheme Regulations.[115] A recent review, however, identified a need for farmers to be better advised on the management options most appropriate for their area and on the effective delivery of these options.[116] In Wales the Tir Cynnal and Tir Gofal schemes both place management obligations upon participating farmers, but Tir Cynnal requires 'levels of environmental protection greater than that of legal and cross-compliance requirements but not as demanding as those of Tir Gofal.'[117]

Farmers in England can also apply to join the scheme at a higher tier, Higher Level Stewardship, receiving higher payments in return for undertaking additional management options.[118] The Countryside Management Scheme in Northern Ireland and the Tir Gofal scheme in Wales enable farmers to do so as well. In Northern Ireland, one of these management options is organic farm management, along with others linked to biodiversity and water quality.

Both the Policy Commission on the Future of Farming and Food and the House of Commons Agriculture Committee, before it, highlighted the need for better financial resources.[119] In practice, financial provision for agri-environment measures has risen considerably. In 1997–1998 £59 million was allocated for agri-environment spending in England.[120] In contrast, £3 billion has been allocated for agri-environment payments in England under the 2007–2013 rural development plan.[121] In tandem with the increased resources and the move away from a concentration of resources within ESA areas, there has also been a marked increase in the amount of farmland enrolled within agri-environment schemes. The area of farmland enrolled in agri-environmental schemes in England at the end of 2005 was six times greater than had been enrolled in 1992.[122] Over 1 million

---

113   The Environmental Stewardship (England) Regulations 2005, Schedule 2 Part 2.

114   *Ibid.* Schedule 3, Part 1.

115   *Ibid.* Regulation 5(9), 5(10) and Schedule 3, Part 4.

116   DEFRA, Natural England, (2008), *Environmental Stewardship: Review of Progress* (London: DEFRA & Natural England) 87.

117   Preamble to the Tir Cynnal (Wales) Regulations 2006.

118   Environmental Stewardship (England) Regulations 2005, Schedule 2, Part 3, which sets out 110 Higher Stewardship management options.

119   See Policy Committee on the Future of Farming and Food, *Farming and Food: A Sustainable Future* (London: The Cabinet Office) 84 and House of Commons Agriculture Committee, footnote 79 above, para. 51.

120   House of Commons Agriculture Committee, footnote 80 above, para. 51.

121   DEFRA webpage, at http://www.defra.gov.uk/rural/rdpe/index.htm. [last accessed 6 August 2009].

122   See DEFRA (2008), *Environment in Your Pocket: 2007* (London: DEFRA) 57.

hectares of land was participating in either the ESA scheme or the Countryside Stewardship scheme in 2005.[123] In contrast approximately 4.7 million hectares is presently participating in the Environmental Stewardship scheme.[124]

---

123 *Ibid.*
124 DEFRA web-page at http://www.defra.gov.uk/rural/rdpe/index.htm. [last accessed 6 August 2009].

Chapter 6
Organic Farming

The Organic Farming Sector

In the 1980s, the European Parliament and European Commission began highlighting the potential role of organic farming within European agriculture.[1] At that stage, the pressing problems were surplus production, agriculture's cost to the European budget and ongoing rural depopulation.[2] Organic farming promoted less intensive production methods and, therefore, resulted in lower production levels. It also attracted higher market prices for producers and was much more labour intensive, thus supporting increased rural employment.[3] Evidence suggested that organic farming's less intensive methods would help to address the environmental problems associated with intensive farming.[4] It also offered a solution to growing consumer demand for high quality food produce produced in a way that not only promoted environmental protection, but also supported animal welfare.[5]

The European Community, therefore, had several objectives for promoting organic farming. At farm level there has been rapid growth in organic farming over the last two decades. In 1985, less than 0.1 per cent of the European Community's agricultural area was being farmed in accordance with certified organic production techniques.[6] By 1999, this had risen to just under 2 per cent,[7] whilst today some

---

1   Lynggard, K. (2008), *The Common Agricultural Policy and Organic Farming: An Institutional Perspective on Continuity and Change* (Wallingford: CAC International) 102.

2   *Ibid.*

3   In relation to this employment issue see Padel, S. Lampkin, N., Farmland Performance of Organic Systems, in Lampkin, N., Padel, S. (eds) (1994), *The Economics of Organic Farming: An International Perspective* (Wallingford: CAB International).

4   See, for example, Dabbert, S., Organic Agriculture and Sustainability: Environmental Aspects, Vetterli, W. et al., Organic Farming and Nature Conservation and Bartram, H., Perkins, A., The Biodiversity benefits of organic farming, all in OECD (2003), *Organic Agriculture: Sustainability, Markets and Policies* (Wallingford: CABI).

5   See generally, Lynggard, K., footnote 1 above, 150.

6   Lampkin, N., Padel, S., Foster, C. Organic Farming, in Brouwer, F., Lowe, P. (eds) (2000), *CAP Regimes and the European Countryside* (Wallingford: CAB International) 222.

7   European Commission (1999), *Agriculture, Environment, Rural Development: Facts and Figures – A Challenge for Agriculture* (Luxembourg: Office for Official Publications of the European Communities) 109.

4 per cent of the agricultural area of the EU-15 is farmed organically.[8] Within the United Kingdom, 4 per cent of agricultural land is also currently being farmed organically,[9] a major increase on only 6,000 hectares in 1985.[10]

Organic production has become increasingly popular amongst consumers seeking high quality produce that is ethically produced through healthy and sustainable methods. This is emphasised by figures produced by the Soil Association, which indicate that the global market for organic produce was worth some £16.7 billion in 2005, with retail sales of organic products in the United Kingdom, that year, being worth some £1.6 million.[11] Despite this growth in organic production, the availability of organic produce still does not match consumer demand. One report in the United Kingdom noted, in 2007, 'substantial domestic undersupply is an issue for organic milk, pork, beef and some fruits and vegetables.'[12] Clearly, there remains potential for further expansion of organic production.

Regulating Organic Production: The Role of Regulation 834/2007

Organic farming began to become more popular in the 1970s and 1980s, however, in the absence of uniform standards, several different approaches to organic farming evolved.[13] This caused consumer confusion, which hindered the development of the sector. This was compounded by the absence of regulations concerning food labelling.[14] The European Community sought to address these issues in 1991, by adopting Regulation 2092/91 on production standards and labelling requirements for organic produce.[15] This Regulation was subsequently amended on numerous occasions, most notably by Regulation 1804/1999. Regulation 2092/91 originally applied only to organic crops and foods based upon those crops. Regulation 1804/1999 extended its scope to include organic livestock products.[16] In 2004, the European Commission published its Action Plan for Organic Food and Farming, calling for the Community to improve and reinforce organic farming standards

---

8   European Commission (2005), *Organic Farming in the EU: Facts and Figures* (Brussels: European Commission) 10. The equivalent figure for the EU-25 is 3.6 per cent.

9   Gill, E. *Organic Farming Under the Green Spotlight*, [2007] 386 ENDS Report 37.

10   Lampkin, N., Padel, S., Foster, C. footnote 6 above, 223.

11   The Soil Association (2006), *Organic Market Report 2006* (Bristol: The Soil Association) 2.

12   E. Gill, footnote 9 above, 37.

13   See European Commission (2001), *Organic Farming: A Guide to Community Rules* (Luxembourg: Office for Official Publications of the European Communities) 4.

14   *Ibid.* 5.

15   Council Regulation 2092/91 [1991] OJ L198/1.

16   Council Regulation 1804/99 [1999] OJ L222/1.

and import and inspection requirements.[17] This was endorsed by the Council of the European Union, which asked the Commission to review the Community's organic farming legislation.[18] The outcome of this review was that Regulation 2092/91 was repealed and was replaced by Regulation 834/2007, on organic production and labelling of organic products.[19] The new Regulation took effect from 1 January 2009.[20] Regulation 834/2007 aims to provide a more explicit statement of the objectives, principles and production rules applying to organic farming and produce. It also introduces new requirements, such as a compulsory European Community logo for organic produce. The Regulation applies to all stages of production, preparation and distribution of organic products, their control and also the labelling and advertising of organic products.[21] It defines organic products as being live or unprocessed agricultural products, processed agricultural products for use as food, feed and vegetative propagating material and seeds for cultivation which are placed upon the market or intended to be placed upon the market.[22]

General Principles and Production Rules for Organic Farming

Regulation 834/2007 establishes three overriding objectives for organic production measures: establishing a sustainable management system for agriculture; producing high quality products; producing a wide variety of foods and agricultural products, which respond to consumer demand for foods produced by methods that do not harm the environment, human health or plant health and which respect animal health and welfare.[23] The Regulation also provides that organic production measures must be based upon the following principles:[24]

a. The appropriate design and management of biological processes based upon ecological systems using natural resources, which are internal to the system, by methods that:

---

17   European Commission, European Action Plan for Organic Food and Farming, COM (2004) 415 final 3.

18   See paragraph 4 of the preamble to Council Regulation 834/2007, [2007] OJ L189/1.

19   Council Regulation 834/2007 [2007] OJ 189/1.

20   *Ibid.* Article 42.

21   *Ibid.* Article 1(1).

22   *Ibid.* Article 1(2). Agriculture in this context also includes aquaculture.

23   *Ibid.* Article 3. The Article, defines a sustainable management system as one that, 'respects nature's systems and cycles and sustains and enhances the health of the soil, water, plants and animals and the balance between them; contributes to a high level of biodiversity; makes responsible use of energy and natural resources such as water, soil, organic matter and air; respects high animal welfare standards and, in particular, meets animals' species specific behavioural needs.'

24   *Ibid.* Article 4.

    i)   Use living organisms and mechanical production methods
    ii)   Practice land-related crop cultivation and livestock production or practice aquaculture that complies with the principle of sustainable exploitation of fisheries
    iii)   Excludes the use of GMOs and products produced from or by GMOs with the exception of veterinary products
    iv)   Are based upon risk assessment, and the use of precautionary and preventative measures, when appropriate

b. The restriction of the use of external inputs
    i)   Where external inputs are required or the appropriate management practices and methods referred to in paragraph (a) do not exist, these shall be limited to:
    ii)   Inputs from organic production
    iii)   Natural or naturally derived substances
    iv)   Low solubility mineral fertilisers

c. The strict limitation of the use of chemically synthesised inputs to exceptional cases these being:
    i)   Where the appropriate management practices do not exist and
    ii)   the external inputs referred to in paragraph (b) are not available on the market
    iii)   Or where the use of external inputs referred to in paragraph (b) contributes to unacceptable environmental impacts

d. The adaptation, where necessary, and within the framework of this Regulation, of the rules of organic production taking account of sanitary status, regional differences in climate and local conditions, stages of development and specific husbandry practices

The Regulation provides that its provisions should apply across the entire agricultural holding.[25] It prohibits organic producers and those, in conversion, from using Genetically Modified Organisms (GMOs) or products produced from GMOs.[26] In applying this provision, producers can reply upon the fact that food or feed is not labelled as being made from GMOs or products produced from GMOs

---

25   *Ibid.* Article 11, though the Article does authorise the European Union to lay down conditions under which holdings can be split into clearly separated units which are not all managed organically.

26   *Ibid.* Article 9, whether as food, feed processing aids, plant protection products, fertilisers, soil conditioners, seeds, vegetative propagating material, micro–organisms and animals in organic production.

or accompanied by documentation to this effect,[27] unless they are aware that the labels do not comply with European Community law.[28]

Member States are required to designate at least one competent authority as the governmental authority responsible for organising the controls required by the Regulation.[29] This authority can authorise independent private bodies to be control bodies tasked with carrying out inspection and certification duties in relation to organic production.[30] Member States are also required to allocate a code number to each control authority and control body.[31] This code number will then be printed on food labelling.[32] In the United Kingdom, the Department of the Environment, Food and Rural Affairs (DEFRA) is the designated competent authority.[33] It is assisted, in this role, by an Advisory Committee on Organic Standards, a non-executive and non-departmental body that advises on organic standards and on the approval of control bodies. Currently, nine private bodies are recognised as control bodies within the United Kingdom.[34] DEFRA also publishes a Compendium of UK Organic Standards, setting out in detail the organic standards required of UK farmers.[35]

Anyone producing, storing or importing organic products must notify this to the competent authority established within their Member State.[36] They must also agree to submit to the control system that is in place within that Member State.[37] For Member States, the Regulation introduced the requirement that the number and frequency of these controls, which will principally be inspections,

---

27   As required under European Union legislation on GMO products – such as Council Directive 2001/18/EC on the deliberate release of GMOs, [2001] OJ L106/1; Council and Parliament Regulation 1829/2003 on genetically modified food and feed, [2003] OJ L268/1; Council and Parliament Regulation 1830/2003 on the traceability and labelling of GMOs and the traceability of products produced from GMOs, [2003] OJ L268/24. See Chapter 9 for an examination of these measures.

28   Council Regulation 834/2007, Article 9(2).

29   *Ibid.* Article 27(1).

30   *Ibid.* Article 27(4) and Article 2(o).

31   *Ibid.* Article 27(10).

32   *Ibid.* Article 24(1).

33   See, the Organic Products Regulations 2004, SI2004/1604.

34   See DEFRA Website, http://www.defra.gov.uk/farm/organic/standards/certbodies/approved.htm [last accessed 6 August 2009].The bodies concerned are the Organic Farmers and Growers Ltd (Certification Code: UK2), The Scottish Organic Producers Association (UK3), The Organic Food Federation (UK4), The Soil Association Certification Ltd (UK5), The Bio-dynamic Agricultural Association (UK6), The Irish Organic Farmers and Growers Association (UK7), The Organic Trust Ltd (UK9), Quality Welsh Food Certification (UK13), Ascisco Ltd (UK15).

35   The compendium may be viewed at http://www.defra.gov.uk/farm/organic/standards/index.htm [last accessed 6 August 2009].

36   Council Regulation 834/2007, Article 28(1).

37   *Ibid.*

are to be determined based on a risk assessment, which considers the risk of irregularities and infringements of the Regulation.[38] Where an irregularity is found, in the way that the organic farming rules are being applied, then, unless this would be disproportionate to the relevance of the requirement breached and to the nature and circumstances of the irregular activities, none of the entire lot or production run affected by that irregularity can be marketed as being organic produce.[39] Where a severe infringement is identified, or an infringement that is likely to have prolonged effects, then the inspection authority must prohibit that producer from marketing produce as being organic for a set period determined by the Member State's competent authority.[40] In contrast, where organic producers and their products meet the requirements of the Regulation then those products can move freely throughout the European Community.[41] Member States can, however, maintain national rules on organic plant or livestock production that are stricter than those introduced by the Regulation, provided those rules also apply to non-organic production and do not prohibit or restrict the marketing of organic products produced outside that Member State.[42]

The Regulation also makes provision for the import of organic produce from third countries.[43] Organic products can be imported from third countries and placed upon the Community market, as long as the product complies with the provisions of the Regulation. Alternatively, organic products may also be imported if they have been produced in accordance with rules that are recognised by the European Community as being the equivalent of those set down in the Regulation, in relation to organic production and the labelling of organic produce. It is, also, an additional requirement that all operators responsible for the production and export of the product have been subject to a control system recognised by the European Commission and that documentary evidence establishes the operators' compliance with these requirements.

Rules Governing Organic Plant Production

To be eligible for organic status, crops must be grown in accordance with plant production rules set out in the Regulation.[44] Conventional farmers must undergo a conversion period, which starts from the time that the producer notifies the competent national authority of his intention to undertake organic production.

---

38   *Ibid.* Article 27(3).
39   *Ibid.* Regulation 30(1).
40   *Ibid.* Article 30(1).
41   *Ibid.* Article 34(1).
42   *Ibid.* Article 34(2).
43   *Ibid.* Articles 32 and 33 (Article 32 concerns the import of products which comply with the requirements of the Regulation, Article 33 concerns the import of products providing equivalent guarantees).
44   Article 6.

The Regulation provides for the length of the conversion period to be defined in subsequent European Community legislation for particular crops and types of animal production. At the time of writing this has not yet occurred. Previously, Regulation 2092/91 stipulated a minimum conversion period of two years (or three years before first harvest in the case of perennial crops) before sown arable crops could be considered to have been produced organically.[45] Similarly, this two-year period also applied before grassland could be harvested as organic feedstuff.[46]

Article 12 of Regulation 834/2007 sets out a number of additional plant production rules. It requires farmers to use tillage and cultivation practices that maintain or increase soil organic matter, enhance soil stability and soil biodiversity and prevent soil compaction and soil erosion. Soil fertility and biological activity is to be maintained and increased by using multi-annual crop rotation patterns, including legumes and other green manure crops, and by applying livestock manure or organic materials from organic production, both preferably composted. Farmers are also required to use production methods that prevent or minimise any contribution to environmental contamination. Plants are to have been derived from seeds, vegetative propagating materials or from plants grown in accordance with the Regulation for at least one generation, or, in the case of perennial crops, for two growing seasons. Damage by pests, diseases and weed is primarily to be prevented through the protection provided by natural enemies, by choice of species and varieties and by the use of crop rotation, cultivation techniques and thermal practices. Although biodynamic preparations may be applied to plants, no mineral fertilisers can be used and other fertilisers, plant protection products and products for cleaning and disinfecting plants during plant production can only be used if they have previously been approved by the European Commission for use in organic production.

Rules Governing Organic Livestock Production

Regulation 834/2007 also sets out specific requirements for organic livestock producers. As in the case of arable farmers, organic livestock producers must undergo a conversion period before they attain organic status. As noted above, the Regulation provides for the length of the conversion periods to be defined in subsequent European Community legislation for specific types of animal production. At the time of writing this had not occurred. Previously, Regulation 2092/91 required farmland used for organic livestock production to go through an analogous conversion period as land used to grow organic crops.[47] A conversion period also applied to livestock themselves. For cattle, used for meat production,

---

45   Council Regulation 2092/91, Annex 1, paragraph 1.1

46   *Ibid.*

47   *Ibid.* Though, by way of exception, Principle 2.1.2 provides that the conversion period for land used by non-herbivore species may be reduced to 1 year and even to 6 months if, in the latter case, it has not received treatments with products other than those

this was 12 months and at least three quarters of the life of the cattle, for pigs and cattle, used for milk production, this was six months. Elsewhere, poultry used for meat production underwent a 10-week conversion period, provided the birds were brought in before they were 3 days old, whilst poultry used for egg production had a six-week conversion period.

Article 14 of Regulation 834/2007 also stipulates further livestock production rules. These are divided into six topic areas: the origin of animals; husbandry practices and housing conditions; breeding; feed and disease prevention and veterinary treatment; the cleaning and disinfection of livestock buildings and installations.

The Origin of Animals   The Regulation requires that organic livestock should be born and raised on organic holdings. However, non-organically raised animals can be brought onto the holding for breeding purposes under specific conditions. These animals and their products may be deemed to be organic after the end of the relevant conversion period for the particular livestock category. Similarly, animals present on the holding at the beginning of the conversion period will be considered to be organic after the conversion period for their particular livestock category has ended.

Husbandry Practices and Housing Conditions

The Regulation addresses issues of animal welfare, environmental protection and the separation of organic and non-organic livestock. In terms of animal welfare, farmers keeping animals are required to have basic knowledge and skills in relation to the health and welfare of animals. Their husbandry, including stock densities and housing conditions, must ensure that animals' developmental, physiological and ethological needs are met. The livestock must have permanent access to open air areas, preferably pasture, when weather conditions and the state of the ground allow – subject to restrictions predicated by the need to protect human or animal health under European Community legislation. The tethering or isolation of livestock is prohibited, unless of individual animals for limited periods where this is justified by safety, welfare or veterinary factors. The duration of animal transport is to be minimised and any suffering during the life of the animal, including mutilation and slaughter, kept to a minimum.

From an environmental perspective, organic livestock farmers are also required to limit the number of livestock that they farm in order to minimise overgrazing, poaching of the soil, erosion or pollution caused either by the animals or by the spreading of their manure.

Finally, in relation to separation, organic livestock must be kept separate from other livestock, though the Regulation does state that the joint grazing of common

---

listed in Annex II of the Regulation. This derogation has to be approved by the organic inspection body.

land by organic and non-organic livestock is permitted under certain restrictive, though unstipulated, conditions.[48]

Breeding Animal breeding is to be based upon natural methods. Although artificial insemination is allowed, other forms of artificial reproduction, such as cloning or embryo transfer, are not. Induced reproduction through hormone treatment, or the use of similar substances, is not allowed unless as a form of veterinary therapeutic treatment for an individual animal. Breeds are to be chosen so as to contribute towards the prevention of suffering and avoiding the need for mutilation of animals.

Animal Feed Animal feed should, primarily, be obtained from the holding upon which the animals are kept or from other organic holdings in the region. The organic feed must meet the animal's nutritional requirements at the various stages of its development. Suckling animals must be fed with natural, preferably maternal, milk and livestock (except bees) must have permanent access to pasture or roughage. A part of each animal's ration may contain feed from holdings in conversion to organic farming but other non-organic feed materials and feed additives can only be used if they have been approved for use in organic production by the European Commission. The use of growth promoters and synthetic amino acids is prohibited.

Disease Prevention and Veterinary Treatment Disease prevention must be based upon the selection of breeds and strains of livestock, husbandry management practices, high quality feed and exercise, appropriate stocking densities and adequate and appropriate housing maintained in hygienic conditions.

Where disease occurs, it must be treated immediately to avoid suffering to the animal. Chemically synthesised allopathic veterinary medicinal products, including antibiotics, can be used where this is necessary and under strict conditions, when the use of phytotherapeutic homeopathic and other products is inappropriate. Restrictions upon courses of treatment and withdrawal periods must be defined where such medicinal products are used. Immunological veterinary medicines can be used. Treatments based upon the protection of human and animal health imposed on the basis of European Community legislation can be conducted.

Cleaning and Disinfection of Livestock Buildings and Installations Only products approved by the European Commission for use in organic production should be used.

---

48 Article 14 also goes on to stipulate husbandry and housing conditions for bees which are not considered here.

Labelling and Advertising Organic Produce

European Community measures on labelling and advertising play a vital role in helping to develop public confidence in organic produce and in combating the fraudulent sale of produce as being organic. In the United Kingdom, these Community measures are enforced by local authorities.[49] Regulation 834/2007 requires the labels of organic produce to set out the code number allocated to the control body that regulates the producer.[50] Additionally, it provides that live or unprocessed agricultural produce can only be described as being 'organic' if all the ingredients of that product have been produced in accordance with the requirements laid down by that Regulation.[51]

In relation to processed food products, Regulation 834/2007, similarly, provides that they can only be described as being 'organic' if they complied with the organic production methods prescribed by the Regulation.[52] The Regulation distinguishes between sales description of the product and its list of ingredients. In order for the word 'organic' to be used in a sales description, the production process of that food must firstly comply with the Regulation.[53] This entails meeting the following stipulations:[54]

a. The preparation of organic and non-organic foods are kept separate in time or space
b. That the product is produced mainly from ingredients of agricultural origin
c. Non-organic agricultural ingredients are only used if authorised by the European Commission or provisionally authorised by a Member State for use in organic production
d. That only additives authorised by the European Commission for use in organic production have been used
e. Organic ingredients are not present with the same ingredient in non-organic form or with an ingredient that was in conversion
f. Food produced from in conversion crops contain only one crop ingredient of agricultural origin

Additionally, at least 95 per cent by weight of the ingredients of agricultural origin of the processed food must be organic.[55]

---

49 Under the Organic Production Regulations 2004, SI 2004/1604, Regulation 10.
50 Council Regulation 834/2007, Article 24(1) (a).
51 *Ibid.* Article 23(1).
52 *Ibid.* Article 23(4).
53 *Ibid.* Article 23(4).
54 *Ibid.* Article 19(1) and (2).
55 *Ibid.* Article 23(4).

To include the word 'organic' in the ingredients list, the production process for that food must again comply with the Regulation. In this instance, this process must comply with stipulations (a), (b), (d) and (e).

The European Community, previously, allowed food products to be labelled as being organic even though they contained genetically modified ingredients. This was confirmed by the European Court of Justice.[56] The Court upheld the validity of Commission Regulation 207/1993, which enabled genetically modified organisms to be included on the list of permitted ingredients of non-agricultural origin that could be contained within food products labelled and advertised as being organic. As one commentator has observed, given consumer resistance to GMO in food, this was a situation that threatened to undermine consumer confidence in organic foods.[57] The European Community has since acted to reverse this situation. Regulation 1804/1999 amended Article 5(3) of Regulation 2092/91 to provide that food products can only be labelled and advertised as being organic if they have been made without the use of genetically modified organisms or of any products derived from genetically modified organisms. As noted above, Regulation 834/2007 today includes a wide-ranging prohibition on the use of gmo materials in organic food and in all stages of organic agricultural production. Additionally, for the purposes of this prohibition, the Regulation also provides that producers and food companies can rely upon the labelling requirements within wider European Community law on gmos. Under the Regulation on Genetically Modified Food and Feed, only produce or products with a gmo content of 0.9 per cent of a food or individual food ingredient need not carry a warning that they contain gmos, as long as this is 'adventitious' or 'technically unavoidable.'[58]

In effect, therefore, a ceiling of 0.9 per cent accidental presence of gmo materials within organic food now applies.

Regulation 834/2007 has also sought to make organic products more recognisable for consumers by introducing a mandatory European Union logo.

Previously, producers had discretion to use the organic farming logo,[59] but from 1 January 2009, it has been compulsory for the Union logo to appear in a conspicuous place on all pre-packaged EU produce that is described as being organic.[60] Products imported from outside the European Union also have discretion to use the logo.[61] The logo must also incorporate an indication of the location at which the agricultural raw materials were farmed. This states 'EU Agriculture',

---

56   See Case C-156/93 *European Parliament v European Commission* [1995] ECR I-2019.

57   Macmaolain, C. (2006), *EU Food Law*, (Oxford: Hart Publishing) 260.

58   Council and Parliament Regulation 1829/2003 on Genetically Modified Food and Feed, [2003] OJ L268/1, Article 12(2).

59   The European Union's organic produce logo was introduced by Commission Regulation 331/2000, [2000] OJ L48/1.

60   Council Regulation 834/2007, Article 24(1).

61   *Ibid.*

where the agricultural raw material was farmed in the European Union or 'non-EU Agriculture' where it was farmed in third countries.[62] Alternatively, it might state 'EU/non-EU Agriculture' if part of the agricultural raw material was farmed in the EU and part in a third country. It is also possible to replace the references to 'EU' and 'non-EU' with a reference to a specific country in which all the agricultural raw materials, of which a product is composed, have been farmed, or to refer to this country in addition to using the 'EU' 'non-EU' indicators.[63] National or private logos can also be used as part of the labelling, presentation and advertising of organic products.[64]

Financial Assistance for Organic Farmers and Farmers Converting to Organic Production

In addition to setting standards for organic production and for the labelling or advertising of organic produce, the European Community has also taken steps to provide financial encouragement and assistance to farmers who convert to organic production. This financial assistance is particularly crucial during the initial conversion period, during which farmers often incur considerable expense in altering their production methods. In its 2004 proposal, for a European Action Plan for organic farming, the European Commission highlighted the need for Member States to make full use of the range of measures available within the rural development programme to support organic farming.[65] This illustrates that no one particular rural development measure is designed to provide specific financial support for organic farming and that this support can come through a range of rural development measures. Today the agri-environment scheme has become a particularly important source of funding for farmers. The environmental benefits associated with organic farming have long been recognised. Indeed when Regulation 2078/92 made it compulsory for all Member States to develop and implement national agri-environment schemes, it specifically cited the fact that agri-environmental schemes could include measures that made payments to farmers who introduced or continued to apply organic farming methods.[66] Today, measures promoting organic farming remain an important element of the agri-environment scheme. Approximately 8 per cent of land enrolled in the scheme

---

62   *Ibid.* Article 24(1). Small quantities, by weight, of ingredients can be disregarded in distinguishing between EU and non-EU farmed produce, provided they do not exceed 2 per cent by weight of all the product's raw materials of agricultural origin.

63   *Ibid.* Article 24(1).

64   *Ibid.* Article 25.

65   European Commission, footnote 18 above, 4.

66   See Council Regulation 2078/92, [1992] OJ L215/85, Article 2(1) (a).

is engaged in organic farming.[67] Equally, the support provided for vocational training and information also plays an important role in helping to finance training and education for producers in organic farming.[68] Similarly, support for organic farming can also be channelled through the rural development's farm modernisation measure, which enables Member States to provide financial assistance for farm investments that improve the economic performance of farms and, somewhat more indirectly, by providing financial support for participation in food quality schemes, to raise consumer awareness of food quality and increase market opportunities for organic produce.[69]

The development of organic farming in the United Kingdom was hindered by previous policies. Payments, here, to farmers converting to organic status were the lowest in the European Community.[70] Additionally, the United Kingdom was one of only four Member States that did not provide on-going financial support to organic farmers after the completion of this conversion period.[71] In contrast, today, farmers throughout the United Kingdom have access to more generous financial payments both during and after conversion. Each UK region, however, structures these payments differently. In England, support for organic farming is provided through 'Organic Entry Level Stewardship', which forms part of England's agri-environment scheme.[72] In Northern Ireland, farmers undergoing conversion to organic status are eligible for payments from a separate organic farming scheme.[73] However, once farmers achieve organic status, subsequent payments are only available through Northern Ireland's agri-environment scheme, Countryside Management. These farmers are eligible to receive higher tier payments in return for agreeing to continue to farm their land in accordance with organic standards.[74] In contrast, in Scotland and Wales, organic farming schemes operate separately from agri-environment measures and make payments to both farmers undergoing

---

67 Based upon the figures supplied in European Commission (2005), *The Agricultural Situation in the European Union, 2004 Report* (Luxembourg: Office for Official Publications of the European Communities) table 3.6.2.3.

68 Council Regulation 1698/2005, [2005] OJ L277/1, Article 21.

69 *Ibid.* Articles 26 and 32 respectively.

70 House of Commons Agriculture Committee (1997), Second Report: *Environmentally Sensitive Areas and Other Schemes Under the Agri-Environment Regulation*, Session 1996–97 (London: HMSO) HC–45–I, xxxi.

71 *Ibid.* xxxii. The others being France, Greece and some German Länder.

72 Under the Environmental Stewardship (England) Regulations 2005, SI2005/621 as amended by the Environmental Stewardship (England) and Countryside Stewardship (Amendment) Regulations 2006, SI 2006/991 and the Environmental Stewardship (England) and Organic Products (Amendment) Regulations 2006, SI2006/2075.

73 Under the Organic Farming Regulations (Northern Ireland) 2008, SR2008/172 as amended by the Organic Farming (Amendment) Regulations (Northern Ireland) 2008, SR2008/435.

74 See Department of Agriculture and Rural Development (2008), *Countryside Management Scheme 2007–2013* (Belfast: DARD) 92.

conversion and those who have achieved that status.[75] Consequently, farmers in Scotland and Wales can participate in both the organic farming schemes and the agri-environment schemes operating in each jurisdiction.

Evaluating the Organic Farming Scheme

Consumers purchase organic products for a variety of reasons. For some, these products form part of a conscious 'natural lifestyle' choice, others are attracted by the health or environmental benefits associated with organic produce, whilst for yet others it may be due to their perception that it looks or tastes better than conventional produce.[76] Whatever the particular motivation, retail sales of organic produce have steadily increased in recent years. In the United Kingdom, retail sales of organic produce virtually doubled between 2000 and 2005.[77]

European Community policy has sought to both promote increased consumer demand for organic produce and to encourage greater numbers of producers to convert to organic methods.[78] Regulation 834/2007 aims to promote consumer demand by making the basic principles of organic production more explicit and, by making the use of the European Union logo compulsory, to increase consumer awareness of organic produce. At a more informal level, Member States have been encouraged to provide better support for organic farming through their rural development policies. The European Community's ultimate objective is to balance both consumer demand and levels of production, so that European agriculture can supply all or most of the consumer demand but, at the same time, avoid the creation of surpluses.

The United Kingdom has no present concern about production surpluses in organic produce. Consumer demand here has consistently outstripped the domestic supply. Indeed, this in turn has led to controversy, with claims that the fuel miles and associated climate change impacts of importing organic produce run counter to the environmental objectives of organic farming.[79] Government policy in the

---

75   See, respectively, the Organic Aid (Scotland) Regulations 2004, SSI2004/143 as amended by the Organic Aid (Scotland) Amendment Regulations 2004, SSI2004/174 and the Organic Aid (Scotland) Amendment Regulations 2005, SSI2005/619 and the Rural Development Programmes (Wales) Regulations 2006, SI2006/3343 (W.304), as amended by the Rural Development Programmes and Agricultural Subsidies and Grant Schemes (Wales) (Amendment) Regulations 2007, SI2007/2900 (W.251).

76   See Lockeretz, W. What are the key issues for consumers?, in OECD (2003), *Organic Agriculture: Sustainability, Markets and Policies* (Wallingford: CAB International Publishing) 241.

77   The Soil Association, footnote 11 above, 46.

78   See, in particular, European Commission, *European Action Plan for Organic Food and Farming*, COM (2004) 415.

79   See, for example, Hickman, M., UK backs African farmers in battle over imported produce, *The Independent*, 26 September 2007, 10.

United Kingdom has sought to increase domestic production levels. For example, the *Action Plan to Develop Organic Food and Farming in England* set a target of raising the percentage of organic produce sold in the United Kingdom, that was domestically produced, from 30 per cent (in 2001/2002) to 70 per cent.[80] This target has almost been achieved; some 66 per cent of organic primary produce sold by multiple retailers in the United Kingdom in 2005 was sourced in the United Kingdom.[81]

Although, initial United Kingdom policy did little to encourage farmers to convert to organic farming, the legislative measures in place today make a more realistic attempt to cover the very real costs associated with conversion to organic status and to support farmers beyond conversion. Despite this, a number of difficulties still persist. Arable farmers, particularly on larger farms, have been reluctant to convert to organic production.[82] This may be due to the fact that conversion entails much greater changes to their farming methods than is the case for other farmers.[83] Equally, organic farming continues to experience 'organic churn', where farmers enter the scheme to access payments for conversion to organic status but then revert to conventional farming once this has been completed. The Soil Association has reported that the area of land farmed organically in the United Kingdom fell by 15 per cent between 2005 and 2006, largely due to the decision of a number of large upland farms in Scotland not to remain in organic production beyond conversion.[84] The fact that such farmers accept payments to convert to organic farming but then do not actually engage in organic production is a waste of the scarce financial resources that the United Kingdom and the European Community have allocated to organic farming. Equally important, is the fact that there is evidence that the amount of land under organic production in the United Kingdom may have gone beyond its peak. The amount of organically managed land (that is fully organic land and land in conversion) in the United Kingdom increased tenfold between 1997 and 2006.[85] However, the 15 per cent fall, noted above, could suggest that this upward trend had now ceased. Indeed, this fall has been part of a general decline in the amount of organically managed land in the United Kingdom from a peak of 726,000 hectares in 2003 to 631,000 hectares in 2006.[86] One explanation is that farmers were responding to uncertainty in the shape of the introduction of the Single Farm Payment and were reluctant to take in-conversion land into organic production at a time when

---

80   Department for Environment, Food and Rural Affairs (2002), *Action Plan to Develop Organic Food and Farming in England* (London: DEFRA) 2.

81   Soil Association, footnote 11 above, 46.

82   Gill, E., footnote 9 above, 35.

83   *Ibid.*

84   Soil Association, footnote 11 above, 17.

85   *Ibid.* 16.

86   *Ibid.*

this new payment regime was being introduced.[87] This theory would seem to be supported by figures showing that the decrease was principally due to a major fall in the amount of land recorded as being in conversion to organic status.[88] In the longer term, the decoupled nature of the Single Farm Payment should actually help to encourage greater support for organic farming. Since the payment is based upon historic production levels, farmers will be able to switch to the lower yields and lower livestock densities associated with organic farming, without affecting the level of Single Farm Payment that they receive. Additionally they will also be able to receive the payments available under the various organic farming schemes operating within each United Kingdom region. In England, this will also engage these farmers within the agri-environment scheme, whilst in Scotland, Wales and Northern Ireland these farmers will have the additional option of taking part in these schemes. These factors may help to explain why the amount of land in conversion to organic status in the United Kingdom actually increased by 68 per cent between 2005 and 2006.[89] Consequently, although the level of organically managed land within the United Kingdom remains below the 2003 peak, it is, perhaps, premature to look upon that year as a watershed.

---

87   See Gill, E., footnote 9 above, 35.

88   Soil Association, footnote 11 above, 17. This notes that where 192,100 hectares were in conversion in 2003 only 58,074 and 51,879 hectares respectively were recorded as being in conversion in 2004 and 2005. In contrast the amount of land under organic production in these three years rose from 534,000 hectares in 2003 to 634,000 in 2005.

89   *Ibid.* 17.

Chapter 7

Agriculture and Nature Conservation

Addressing Agriculture's Role in Protecting Biodiversity

In recent years, continued biodiversity decline has become an issue of global concern. So much so that, in 2002, the 6th conference of the parties to the United Nations Convention on Biological Diversity adopted the target of achieving a significant reduction in rates of biodiversity by 2010.[1] This target was also, subsequently, endorsed by 2002 World Summit on Sustainable Development.

The European Union is no different from other parts of the world. It has also experienced serious biodiversity decline. The European Environment Agency has summarised the findings of a number of studies by conservationist and conservation groups in the following terms:[2]

- 42 per cent of European mammals are threatened at the global level, including the Iberian Lynx and the Mediterranean Monk Seal
- 43 per cent of the European avifauna has an unfavourable conservation status in Europe
- Most fish stocks of commercial importance in European waters appear to be outside safe biological limits
- 12 per cent of the 576 butterfly species resident in Europe are very rare or seriously declining on the continent
- Up to 600 European plant species are considered by the International Union for Conservation of Nature to be extinct, extinct in the wild or critically rare. Of these only half are cultivated in botanical gardens which ensure *ex situ* conservation
- 26.8 per cent of the world's domestic mammalian breeds and 57.6 per cent of poultry breeds currently at risk of extinction occur in Europe
- 22 per cent of the 2,238 European breeds registered by the United Nations Food and Agriculture Organisation have approached a critical population size, 34 per cent are classified as endangered and 44 per cent are considered to be at risk of being lost

---

1 See Convention on Biological Diversity, Decision 6/26: Strategic Plan for the Convention on Biological Diversity (19 April 2002) available at http://www.cbd.int/decision/cop/?d=7200.

2 European Environment Agency (2006), *Progress Towards Halting the Loss of Biodiversity by 2010* (Copenhagen: European Environment Agency) 21.

The European Union is a party to the United Nations Convention on Biological Diversity. However, it has chosen to adopt an even more stringent policy goal than that adopted under the Convention. The EU's Strategy for Sustainable Development, endorsed by the European Council in 2001, provides for it to 'protect and restore habitats and natural systems and *halt* the loss of biodiversity by 2010.'[3] This goal was also repeated, in 2002, within the EU's 6th Environmental Action Programme, which identified the protection of nature and biodiversity as a priority area for environmental policy.[4]

Agriculture, and the Common Agricultural Policy in particular, can have an important role in helping the European Community meet its goal of halting biodiversity loss. Equally, however, it is clear that they have also been part of the problem of biodiversity decline.[5] Agricultural landscapes have become vulnerable to two twin threats – the intensification of farming and the abandonment of farmland. The challenge for the European Community has been to support and encourage those agricultural practices that are beneficial to wildlife, whilst simultaneously discouraging those which have been at the root of population declines, such as those noted amongst farmland birds and butterflies.

The European Community has, essentially, adopted a twin track strategy in seeking to achieve its objective of halting biodiversity loss by 2010. On the one hand, it seeks to provide special protection for sites of particular nature value. On the other hand, it has also recognised that it cannot achieve its objective by merely protecting these 'islands of conservation'. Through the agri-environment scheme it also provides broader measures to protect biodiversity in the countryside as a whole.[6] The Common Agricultural Policy plays an important role in relation to both parts of the strategy. As will be noted below, agricultural policy measures have been introduced to support both the operation of the European Community's nature conservation legislation and also to promote nature conservation measures in the wider countryside. Additionally, agricultural policy has also been given the, perhaps even more fundamental role of providing financial support for measures introduced at farm level.

---

3   See European Commission, *A Sustainable Europe for a Better World: A European Union Strategy for Sustainable Development*, COM (2001)264 final, 12. [Emphasis added by author].

4   Decision No.1600/2002/EC of the European Parliament and of the Council laying down the 6th Environmental Action Programme, [2002] OJ L242/1, Article 6(1).

5   See earlier discussion in Chapter 2.

6   See, for example, European Commission, *Halting the Loss* of *Biodiversity by 2010 – And Beyond: Sustaining Ecosystem Services for Human Well-being*, COM (2006)216 final, 6.

Protecting Important Species and Their Habitats: The Wild Birds and Habitats Directives

As mentioned in the previous paragraph, one part of the European Community strategy for halting biodiversity loss is based upon the protection of sites of particular nature conservation importance. The European Community has adopted two directives that aim to play an important role in delivering protection to such sites:

• Directive 79/409/EEC on the Conservation of Wild Birds (the Wild Birds Directive)[7]
• Directive 92/42/EC on the Conservation of Natural Habitats and of Wild Fauna and Flora (the Habitats Directive)[8]

Collectively these Directives seek to protect a wide range of flora and fauna across the European Union.

In the United Kingdom, it is government policy that all areas that are being considered for designation under either Directive should be, first, designated as an SSSI, under national law. The United Kingdom's nature conservation legislation was, initially, quite weak. Consequently, the Conservation (Natural Habitats etc.) Regulations 1994 were adopted to augment these laws, with a view to implementing the Wild Birds and Habitats Directives.[9] However, the legal protection provided to SSSIs has been strengthened, significantly, in the last decade. In England and Wales, the legal controls upon SSSIs are set out in schedule 9 of the Countryside and Rights of Way Act 2000, which inserted these provisions into the 1981 Wildlife and Countryside Act. In Scotland, they are governed by the Nature Conservation (Scotland) Act 2004 and in Northern Ireland by the Environment (Northern Ireland) Order 2002. Protection for nature conservation sites, identified under the Directives, continues to be provided through a combination of the national legislation regarding SSSIs and the Conservation Regulations. However, the more robust nature of the protection now provided to SSSI sites means that legislation such as the Countryside and Rights Act now plays a more prominent role in protecting these sites. The Conservation Regulations themselves have also been amended,[10] following a decision, by a European Court of Justice, finding

---

7   [1979] OJ L103/1.

8   [1992] OJ L206/7.

9   SI 1994/2716 as amended, in Northern Ireland this was achieved by the Conservation (Natural Habitats etc.) (Northern Ireland) Regulations 1995, SR 1995/380, as amended.

10   See the Conservation (Natural Habitats etc.) (Amendment) (England and Wales) Regulations 2007 and 2009, SIs 2007/1843 and 2009/8, the Conservation (Natural Habitats etc.) Amendment (Scotland) Regulations 2007 and 2008, SSIs 2007/349 and 2008/425 and the Conservation (Natural Habitats etc.) (Amendment) Regulations (Northern Ireland) 2007 and 2009, SRs 2007/345 and 2009/8.

that the existing Regulations had not fully implemented the Habitats Directive.[11] In particular, the Court criticised the general nature of the obligation placed upon government departments to comply with the Habitats Directive, and the fact that the Regulations did not require assessment of the implications for protected sites, of land use or water abstraction plans.[12]

The Wild Birds Directive 1979

The Wild Birds Directive was the first piece of nature conservation legislation to be adopted by the European Community. The Directive deals with a number of issues concerning the protection of wild birds, such as deliberate killing or capture of wild birds, damaging their nests or eggs and trade in live or dead birds. Additionally, it also seeks to protect their habitats. It is this aspect of the Directive that is of interest here.

The Directive places a general obligation upon all Member States to take action to preserve, maintain or re-establish a sufficient diversity and area of habitats of all, naturally occurring, species of wild birds that occur within their European territories.[13] Their objective is to maintain the population of these birds at a level 'which corresponds, in particular, to ecological, scientific and cultural requirements, while taking account of economic and recreational requirements, or to adapt the population of those species to that level.'[14] Ireland, for example, was found to have breached these requirements by failing protect the bog and heather habitats, of the red grouse, from serious degradation through overgrazing by sheep.[15] The Court emphasised the preventative nature of the obligation imposed upon Member States and pointed to evidence of a considerable reduction in the range of the species and of a clear and severe deterioration of its habitat in order to establish Ireland's failure to comply with its obligation under the Directive.[16]

The Wild Birds Directive also requires Member States to implement special conservation measures to protect the habitats of a number of particularly vulnerable species of wild birds, listed in Annex I to the Directive.[17] Member States must classify their most suitable territories, in number and size, as being 'Special Protection Areas' and 'take appropriate steps to avoid pollution or deterioration of habitats or any disturbances affecting the birds, in so far as these would be significant, having regard to the objectives of this article'.[18] An identical obligation

---

11   Case C-6/04 *European Commission v United Kingdom* [2006] Env. L.R. 29.
12   For an analysis of the judgment see Reid C.T. (2006), Implementing EC Conservation Law, [2006] 18 *Journal of Environmental Law*, 135.
13   Wild Birds Directive 1979, [1979] OJ L103/1, Articles 1 and 3.
14   *Ibid.* Article 2.
15   Case C-117/00 *European Commission v Ireland*, [2002] I-5335.
16   *Ibid.* Para. 70.
17   Wild Birds Directive 1979, Articles 4(1) and 4(2).
18   *Ibid.* Articles 4(1), 4(2) and 4(4).

also applies with respect to the habitats of regularly migrating birds.[19] The Directive here requires that Member States should pay particular attention to the protection of wetlands – especially those of international importance.[20] Currently, in the United Kingdom, 256 special protection areas have been designated, covering a total area of 1,610,812 hectares.[21] An issue of concern for Member States has been the extent to which they are able to balance the designation of special protection areas with broader, human, social and economic needs. Although the Directive allowed Member States to take account of economic and recreational requirements when exercising their general duty to protect the habitats of all naturally occurring species of European wild birds, no such provision was included in relation to Special Protection Areas. In reality the European Court of Justice has adopted a strict approach on this issue. It has ruled that Member States are only entitled to take ornithological criteria into account when considering the designation of Special Protection Areas.[22] Additionally, the Court has accepted ornithological evidence introduced by the European Commission, in the form of the 1989 Inventory of Important Bird Areas in Europe, compiled by Birdlife Europe, to establish that Member States have failed to designate a sufficient number or area of sites as being Special Protection Areas.[23] Member States have also been required to protect areas, that should have been designated as Special Protection Areas, to the same extent as those that have actually been designated.[24] They must implement sufficient measures to ensure the survival and reproduction of species listed in Annex I and also to ensure the breeding, moulting and wintering of migratory birds.[25] The obligation to protect these habitats has been found to arise even before any reduction has been observed in bird numbers or any risk of their extinction has materialised.[26] Elsewhere, the Court also adopted a similarly restrictive approach in relation to proposals, by Member States, to alter the area of designated Special

---

19   *Ibid.* Article 4(2).

20   *Ibid.*

21   See Department for Environment, Food and Rural Affairs webpage http://www. defra.gov.uk/wildlife-countryside/protected-areas/spa.htm [last accessed 2 November 2008].

22   See Case C-355/90 *European Commission v Spain* [1993] ECR I-4221 (Santoña Marshes) and Case C-44/95 *R v Secretary of State for the Environment, ex parte Royal Society for the Protection of Birds* [1996] ECR I-3805 (Lappel Bank). See also Case C-247/85 *European Commission v Belgium* [1987] ECR 3029 and Case 262/85 *European Commission v Italy* [1987] ECR 3073.

23   See Case C-3/96 *European Commission v The Netherlands* [1998] ECR I-3031 and Case C-166/97 *European Commission v France* [1999] ECR I-1719. Note, however, that in *Brown v Secretary of State for Transport* [2004] Env. LR 26, the court did accept evidence that showed that the 1989 study was no longer accurate.

24   See Case C-355/90 *European Commission v Spain* [1993] ECR I-4221 para 57.

25   Under Articles 4(1) and 4(2) of the Wild Birds Directive 1979, see also Case C-166/97 *European Commission v France* [1999] ECR I-1719, para. 22.

26   *Ibid.* para. 15.

Protection Areas for non-ornithological reasons. Member States can only make such amendments to the boundaries of Special Protection Areas, in particular, by reducing them in size, on exceptional grounds that supersede the ecological interests protected by the Directive. [27] One such ground being the protection of human life.[28]

The Habitats Directive 1992

In adopting the Habitats Directive, the European Community acted to extend the scope of EU nature conservation law by incorporating the protection of wild animals and plants within its scope. However, equally, this Directive also acted to moderate the strict approach adopted by the European Court of Justice in interpreting the level of protection required by the Wild Birds Directive.

Like the Wild Birds Directive, the Habitats Directive adopted a two pronged approach. First, it prohibits activities such as the capture or killing of particular animals identified in the Directive and destruction or taking of their eggs.[29] Similarly, activities such as picking, uprooting or destroying particular plant species are also prohibited.[30] Member States must also prevent trade in these animal and plant species.[31]

Second, the Habitats Directive also introduced measures to require Member States to protect particular wild animals and plants and their habitats. However, it did so in a manner that was radically different from the Wild Birds Directive. The stated aim of the Habitats Directive is to 'contribute towards ensuring biodiversity through the conservation of natural habitats and of wild fauna and flora in the European territory of the Member States.'[32] This is to be achieved by introducing measures designed to 'maintain or restore, at a favourable conservation status, natural habitats and species of wild fauna and flora of Community interest.'[33] At the same time, however, the Directive specifically provides that measures taken under it 'shall take account of economic, social and cultural requirements and regional and local characteristics.'[34] Therefore, unlike the Wild Birds Directive, provision has been made for these considerations to be taken into account when measures were being designed to protect fauna and flora, and their habitats identified by the Directive.

---

27   See Case C-57/89 *European Commission v Germany* [1991] ECR I-883 (Leybrucht Dykes).

28   *Ibid.*

29   The Habitats Directive 1992, [1992] OJ L206/7, Article 12.

30   *Ibid.* Article 13.

31   *Ibid.* Articles 12 and 13.

32   *Ibid.* Article 2(1).

33   *Ibid.* Article 2(2).

34   *Ibid.* Article 2(3).

The specific objective of the Habitat's Directive, as set out in Article 3, is the creation of a 'coherent European ecological network of special areas of conservation' to be known as *'Natura 2000'*. This network is to comprise:[35]

1. Sites hosting natural habitats of the type listed in Annex I to the Directive
2. Sites providing habitats for particular species of flora and fauna listed in Annex II to the Directive
3. Special Protection Areas designated under the Wild Birds Directive

In total Annex I to the Habitats Directive lists 168 habitat types, whilst Annex II includes 632 species of wild flora and fauna whose habitats are to be protected. Additionally, Annex I and Annex II identify a number of especially vulnerable habitat types (Annex I) or wild flora and fauna (Annex II) as being a priority. There was no equivalent to this group under the Wild Birds Directive, which did not single out any particular bird species as being conservation priorities.

The method of site designation that applies under the Habitats Directive is also different from that which applied under the Wild Birds Directive. Where the Wild Birds Directive placed the onus on Member States to designate appropriate areas as being Special Protection Areas, the Habitats Directive required Member States to collaborate with European Community institutions, particularly the European Commission. The Habitats Directive, essentially, put in place a four step designation procedure.

Step One   Member States were each required to propose a list of potential sites of Community interest to the European Commission.[36] Annex III to the Directive provides guidelines to assist Member States in identifying sites that either host the natural habitat types listed in Annex I to the Directive or the species listed in Annex II.[37] These sites of potential Community importance were to be transmitted to the European Commission by June 1995.[38]

Step Two   Once the Member States had nominated potential sites to the European Commission, it was then required to establish a draft list of 'sites of Community importance.'[39] Again, Annex III to the Directive provided guidance. For example, all sites, proposed by Member States, that contained priority natural habitat types and/or priority species, were to be considered to be sites of Community importance. The European Commission was required to establish its draft list 'in agreement with each Member State', thus giving Member States a measure of control over its

---

35   *Ibid.* Article 3(1).
36   *Ibid.* Article 4(1).
37   *Ibid.* Guidance is also set out in Article 4(1).
38   *Ibid.* Article 4(1) requires transmission to take place within three years of the notification of the Directive.
39   *Ibid.* Article 4(2).

content. Equally, where a site contained a priority habitat or species, the Directive gave the European Commission a power to propose it as being of Community importance even though it had not been included on the list of sites initially notified by a particular Member State.[40] However, the Directive here left control with Member States. In the absence of the agreement of the Member State in which the site was located, the Commission was required to notify the Council, who could adopt a decision to add the site to the Commission's draft list of sites of Community importance.[41] The Council was required to act unanimously in doing so, thereby providing a veto to the Member State concerned.[42] The draft list of sites of Community importance was to be in place by June 1998.[43]

Step Three   Once the draft list of sites of Community importance was in place, the Commission was required to obtain European Community approval of it.[44] The list was to be referred to a committee, composed of representatives of the Member States, with the Commission acting as, non-voting, chair.[45] If this committee voted, by qualified majority, to accept the draft list then it could be adopted by the Commission.[46] Alternatively, it would have to be referred to the Council, which could vote to adopt it – this time by a qualified majority.[47]

Step Four   Member States were placed under an obligation to designate each site of Community importance as being a Special Area of Conservation.[48] This designation was to be completed within 6 years of the site being identified as being of Community importance under step three.[49]

One issue that arose, in relation to this designation process, was whether Member States could take economic, social or cultural requirements into account when deciding whether to notify potential sites of Community importance to the European Commission? The European Court of Justice has now answered this question, negatively. It has confirmed that only conservation-based issues could be taken into account when deciding which areas should be notified to the European Commission.[50] The Court reached this view on the basis that any other conclusion

---

40   *Ibid.* Article 5.

41   *Ibid.* Article 5(2).

42   *Ibid.* Article 5(3).

43   *Ibid.* Article 4(3), which required that this list be in place within 6 years of the notification of the Directive.

44   *Ibid.* Article 4(2).

45   *Ibid.* Article 4(2) and Articles 20 and 21.

46   *Ibid.* Article 4(2) and Article 21.

47   *Ibid.* Qualified majority voting is a voting system that weights the number of votes given to each Member State in proportion to their relative population size.

48   *Ibid.* Article 4(4).

49   *Ibid.*

50   Case C-371/98 *R v Secretary of State for the Environment, Transport and the Regions, ex parte First Corporate Shipping* [2001] ECR I-9235, paras. 13–16.

would mean that the European Commission would not have had an exhaustive list of eligible sites. It took the view that such a situation could compromise the Directive's objective of creating a coherent European ecological network.[51] However, as noted above the Habitats Directive does authorise Member States to take into account economic, social and cultural requirements, along with regional and local characteristics when designing measures to maintain or restore the habitats and species that it protects.[52] This raises the issue of how such considerations are to be taken into account by Member States.

The protection of Special Areas of Conservation is an area governed by Article 6 of the Habitats Directive. The provisions of this Article have also been copied into the Conservation (Natural Habitats etc.) Regulations 1994 in Great Britain, to set out the issues conservation authorities here should take into account when deciding whether to authorise projects that are proposed by owners or occupiers of European Sites. Article 6 requires Member States to establish 'the necessary conservation measures involving, if need be, appropriate management plans specifically designed for the sites integrated into development plans and appropriate statutory, administrative or contractual measures which correspond to the ecological requirements' of the natural habitat types and species protected within the site.[53]

In similar terms to the demands made by the Wild Birds Directive, Member States are required to 'take appropriate steps to avoid, in the special areas of conservation, the deterioration of natural habitats and the habitats of species as well as disturbance of the species for which the area has been designated, in so far as such disturbance could be significant in relation to the objectives of this Directive.'[54]

More specifically, an environmental impact assessment must be conducted on any plan or project not directly concerned with the conservation management of the site but which is likely to have a significant effect upon it.[55] National authorities can, then, only authorise the plan or project if the assessment shows that it will not adversely affect the integrity of the site.[56] However, this protection is subject to an important limitation. Member States will be entitled to approve plans or projects, despite a negative environmental assessment, where imperative reasons of overriding public interest, including those of a social or economic nature, justify

---

51   *Ibid.* para. 24.

52   Habitats Directive 1992. Article 2(3).

53   *Ibid.* Article 6(1).

54   *Ibid.* Article 6(2).

55   *Ibid.* Article 6(3).

56   *Ibid.* The article also provides that, 'if appropriate', this decision should be made after having obtained the opinion of the general public. This, therefore, leaves Member States to decide upon the level of public consultation that should be incorporated within this procedure.

carrying them out and no alternative solution exists.[57] In this situation Member States must take sufficient compensatory measures to ensure that the overall coherence of the *Natura 2000* network is protected and also inform the European Commission of these measures.[58] Member States, however, enjoy more limited discretion to authorise plans or projects that will damage the conservation interest of a Special Area of Conservation, where the site contains a priority habitat type or priority species.[59] In that case, the only imperative reasons of overriding public interest, that can be taken into account, are those concerned with 'human health, or public safety, beneficial consequences of primary importance for the environment or, further to an opinion from the Commission, other imperative reasons of overriding public interest.'[60] Few Commission opinions have been delivered on this issue. However, those that have show that the Commission has been prepared to authorise projects on economic grounds, whilst taking account of issues such as whether alternatives existed or whether adequate compensation measures were proposed.[61]

The provisions outlined in the previous paragraph represent the Member States attempt to move away from the very strict interpretation adopted by the European Court of Justice in relation to the protection of Special Protection Areas. They still seek to put in place a high level of conservation protection, whilst also ensuring that conservation goals can be overridden, in particular circumstances, in pursuit of other, non-conservation, objectives. These provisions are particularly significant because the Habitats Directive also provides that, from the date of its implementation or the date of designation of the site – whichever is later, they will also apply to Special Protection Areas designated under the Wild Birds Directive.[62] This, therefore, gives more scope for Member States to authorise projects that may damage the conservation interest of Special Protection Areas designated under the Wild Birds Directive.[63] As noted above, the Habitats Directive specifically provides that social and economic factors can justify particular works, something that the European Court of Justice had, previously, excluded under the Wild Birds Directive.[64] The Court, however, has gained some revenge. It has found that these aspects of the Habitats Directive will not apply in relation to the management of sites that ought to have been designated as Special Protection Areas, but have not

---

57   *Ibid.* Article 6(4).

58   *Ibid.* For example, Reid points out that when the Cardiff Bay barrage scheme resulted in the loss of important feeding grounds for wading birds, a new wetland reserve was created elsewhere as a compensatory measure. See Reid, C. (2002), *Nature Conservation Law*, 2nd ed. (Edinburgh: W. Green & Son Ltd), para.5.2.23.

59   *Ibid.* Article 6(4).

60   *Ibid.* Article 6(4).

61   See, for example, Nollkaemper, A., Habitat Protection in the European Community: Evolving Concepts of a Balance of Interests, [1997] *Journal of Environmental Law* 271.

62   *Ibid.* Article 7.

63   In case C-57/89 *European Commission v Germany* [1991] ECR I-883.

64   *Ibid.* para. 22.

been designated.[65] These will continue to be governed by the Wild Birds Directive and by the more restrictive case law developed by the Court under that Directive. However, it is of interest that the Wild Birds Directive does not make provision for priority bird species. Consequently, Special Protection Areas, which have been designated, do not qualify for the enhanced protection provided under the Habitats Directive to sites containing priority habitat types or priority species.

Progress Towards Natura 2000

The Wild Birds and Habitats Directives may seem to put in place a comprehensive legal mechanism through which vulnerable and endangered wildlife can be safeguarded throughout the Community. In reality, Member States' implementation record in relation to each Directive has been less than impressive.

The European Commission recently reported that, almost 30 years after it was adopted, it considered that only four Member States have designated sufficient Special Protection Areas.[66] Even greater problems have arisen in relation to the Habitats Directive. No Member State met the 1995 deadline to nominate potential sites of Community importance to the European Commission.[67] By the end of 2000 the Commission had still only received partial information from Member States.[68] This delayed the Commission and prevented it from making its own selection of sites of Community importance. In reality, the first sites of Community importance were not identified by the Commission until December 2001.[69] Those in the United Kingdom and Ireland were chosen in 2004.[70] Across the European Community as a whole, whilst a large number of sites of Community importance have been chosen, the selection process still remains uncompleted.[71] In the United

---

65   See Case C-374/98 *European Commission v France* [2000] ECR I-10799.

66   European Commission, *Report on the Implementation of Council Directive 79/409/EEC on the Conservation of Wild Birds*, COM (2006)164 final, 9.

67   Holdaway, E., Holdaway, H. (2000), *Making the Habitats Directive Work* (London: WWF) 3.

68   European Commission, *Report on the Implementation of Council Directive 92/43/EC on the Conservation of Natural Habitats and of Wild Fauna and Flora*, COM (2003)845 final, 16.

69   See Commission Decision 2002/11/EC of 28 December 2001 adopting the list of sites of Community importance for the Macronesian biogeographical region, [2002] OJ L5/16.

70   See Commission Decision 2004/813/EC adopting the list of sites of Community importance for the Atlantic biographical region, [2004] OJ L397/1 as amended by Commission Decision 2008/23/EC adopting the first updated list of sites of Community importance for the Atlantic biographical region [2008] OJ L12/1.

71   Under Article 2(1) of the Habitats Directive 1992, as amended, the European Community has been divided into nine biogeograpical regions and sites of Community importance then chosen for each region. By October 2008 sites of Community importance

Kingdom we have designated 608 Special Areas of Conservation, covering a total of 2,505,165 hectares.[72]

The delays that have occurred during the implementation of the Habitats Directive can only have had a negative impact upon biodiversity. This can also be seen from the face of the Directive itself. For example, the Directive requires Member States to protect sites from deterioration or from the damaging impact of plans or projects, from the time that they are chosen by the Community as Sites of Community importance.[73] The European Commission has argued that this obligation should actually be considered to begin earlier, at the date on which Member States nominate potential sites of Community importance to it.[74] The Commission developed this argument by analogy with the Court's, highly protective, case-law under the Wild Birds Directive, where the Court has held that Member States should protect sites that ought to have been designated as Special Protection Areas under that Directive.[75] But, whichever of these two approaches is adopted, delays have occurred in Member States nominating sites and, consequently, in the Commission selecting its sites of Community importance. These delays can only have exposed these sites to the potential damage of their nature conservation interest. If consideration is given, not only, to protecting sites from damage, but, also, to encouraging their enhancement through positive management, then similar issues arise. The Habitats Directive provides that Member States should establish such positive conservation measures for special areas of conservation.[76] This would imply that such measures are only obligatory once sites have been designated as being Special Areas of Conservation under national laws. As has already been observed, Member States have up to six years to take this step, from the time that a site is chosen as a site of Community importance. However, some Member States have opted to support positive management at an earlier stage. In the United Kingdom, potential sites of Community importance are first designated as Sites of Special Scientific Interest under national nature conservation laws.[77] This has made positive management measures available, under these national nature conservation laws, from the time that sites were nominated to the European

---

had not yet been chosen for one of these regions – the Black Sea region. For details of the measures adopted by the Commission in order to specify sites of Community importance in each of the other biogeographical regions see http://www.ec.europa.eu/environment/nature/home.htm [last accessed 6 August 2009]

72   See Department of Environment, Food and Rural Affairs website http://www.defra.gov.uk/wildlife-countryside/protected-areas/sac.htm [last accessed 2 November 2008].

73   Habitats Directive 1992, Articles 4(5) and 6(2)–6(4).

74   European Commission (2000), *Managing Natura 2000 Sites: The Provisions of Article 6 of the Habitats Directive 92/43/EEC* (Brussels: European Commission) 13.

75   Case C-355/90 *European Commission v Spain* [1993] ECR I-4221, para. 57.

76   Habitats Directive 1992, Article 6(1).

77   Known as Areas of Special Scientific Interest in Northern Ireland.

Commission.[78] Some other Member States, such as Portugal, Greece and parts of Spain, have introduced management measures, when sites have been chosen by the Commission as sites of Community importance.[79] However, once again, the delays that have occurred at each stage of the adoption of these sites, under the Directive, can only have had a negative impact for biodiversity protection across the European Union.

CAP's Role in Supporting the Natura 2000 Network

There can be no doubt that the CAP has a vital role to play in supporting the European Community's goal of halting biodiversity loss. The importance of this role has been recognised by the Community. At a general level, biodiversity protection is one of the key themes identified in the European Community's 6th Environmental Action Programme. More particularly in relation to agricultural policy, the 2008 Health Check reform proposals acknowledged that the CAP needed to do more to protect biodiversity.[80] The European Commission also highlighted agriculture's potential role when it published a Biodiversity Action Plan for Agriculture.[81] This, essentially, provided for the Community to ensure that biodiversity protection was adequately integrated into general CAP measures. As the European Commission itself has commented, the priorities for agriculture set out in this plan were:[82]

> ...the promotion and support of environmentally friendly farming practices and systems which benefit biodiversity directly or indirectly, the support of sustainable farming activities in biodiversity-rich areas, the maintenance and enhancement of good ecological infrastructures, the promotion of measures related to genetic resources, and the development of the marketing of non-commercial varieties.

One important aspect of agriculture's role is the support that it can provide for the operation of the Wild Birds Directive and Habitats Directive. This is illustrated by the fact that agricultural habitats account for some 35 per cent of the

---

78 European Commission, *Report on the Implementation of Directive 92/43 on the Conservation of Natural Habitats and of Wild Fauna and Flora*, COM (2003)845 final, 29.

79 *Ibid.*

80 European Commission, *Preparing for the 'Health Check' of the CAP Reform*, COM (2007)722, 9.

81 European Commission, *Biodiversity Action Plan for Agriculture*, COM (2001)162 final.

82 European Commission, (2004), *Biodiversity Plan for Agriculture: Implementation Report* (Brussels: (European Commission) 4.

total area proposed as potential sites of Community importance by the EU-15.[83] Similarly, it has been estimated that 65 of the 198 habitat types whose protection is required under the Habitats Directive are threatened by agricultural intensification, whilst a further 32 habitat types are threatened by the abandonment of extensive grazing.[84]

As was noted above, Member States are required take appropriate action to avoid deterioration of natural habitats and of the habitats of protected species. The cross-compliance measures, now included within CAP's production policies, provide one mechanism through which the European Community has been able to assist Member States in meeting this obligation. The Common Agricultural Policy's current cross-compliance requirements replace more liberal measures that had previously existed. In truth, these earlier measures only provided limited protection for wildlife. In the case of the direct payments made to farmers as part of production policy, Member States had simply been required to 'take the environmental measures they consider to be appropriate in view of the situation of the agricultural land used or the production concerned and which reflect the potential environmental effects.'[85] Eight Member States, from the EU-15, responded to this provision by introducing cross-compliance measures that made direct payments conditional upon environmental requirements.[86] However, the European Commission has reported that only two, Finland and Ireland, have established cross-compliance requirements that specifically aim to protect biodiversity.[87] In contrast, the less favoured area scheme previously required farmers to apply 'usual good farming practices compatible with the need to safeguard the environment and maintain the countryside',[88] whilst the environmentally sensitive areas scheme had required farmers to go beyond 'usual good farming practice'.[89] In the absence of any Community law framework, Member States developed a wide variety of approaches to define good farming practice. One criticism that has been made is that these often paid limited attention to practices that had a direct impact upon nature conservation.[90] Another criticism is that some contained few verifiable standards, making their enforcement difficult.[91]

Today, as was noted in Chapter 3, the Wild Birds Directive and Habitats Directive are both included in the list of Community legislative measures, with

---

83   European Environment Agency (2006), *Progress Towards Halting the Loss of Biodiversity by 2010* (Copenhagen: European Environment Agency) 42.

84   *Ibid.*

85   Council Regulation 1259/99, [1999] OJ 160/113, Article 3.

86   European Commission (2004), Working Document: *Biodiversity Action Plan for Agriculture – Implementation Report* (Brussels: European Commission,) 20.

87   *Ibid.*

88   Council Regulation 1257/99, [1999] OJ L160/80, Article 14(2).

89   *Ibid.* Article 23(2).

90   European Commission, footnote 86 above, 20.

91   *Ibid.* 19.

which farmers must comply in order to remain eligible to receive Single Farm Payments. Farmers, therefore, have a direct incentive to protect Natura 2000 sites. Those who fail to do so face the possibility of having their payments reduced and, subsequently, removed. The scope of this cross-compliance requirement has also been extended by Regulation 1698/2005, so that farmers, receiving payments under the less favoured area payments or under the European Community's Environmentally Sensitive Areas scheme, are also now specifically required to comply with the Wild Birds Directive and Habitats Directive. However, these measures will only be effective if Member States, themselves, take adequate enforcement measures. As highlighted in Chapter 3, Member States have, traditionally, been reluctant to support cross-compliance measures. An important question mark must, therefore, remain as to their enthusiasm for enforcing these measures. The European Commission will, therefore, need to ensure that Member States do take effective action. Equally, it should also be recalled that cross-compliance can only provide a partial answer. It will have no impact upon agricultural sectors such as pig or poultry farming, where no Single Farm Payments are made, unless farmers within these sectors are participating in either the less favoured area or environmentally sensitive areas schemes.

The CAP has an equally important role in helping Member States to meet their obligation, to maintain protected habitats and species at favourable conservation levels.[92] In particular, it can assist them in fulfilling the Directive's requirement that they should implement 'the necessary conservation measures' that 'correspond to the ecological requirements' of these habitats and species.[93] One major issue that arises here is the significant cost involved in providing positive management on sites protected under the Wild Birds and Habitats Directives. The European Commission has estimated that the cost of managing the Natura 2000 network, as a whole, will vary between €3.4 billion and €6.1 billion per year over the period 2003 to 2013.[94] In practice the management costs involved in maintaining Natura 2000 sites have been identified as one of the factors that encouraged Member States' to delay the notification of potential Natura 2000 sites.[95]

Article 8 of the Habitats Directive does authorise the European Union to provide co-funding for conservation measures in Natura 2000 sites that contain priority habitat types or species. Member States can ask the European Union to consider co-funding measures that are essential for the conservation of one or more priority habitat or species. However, the European Community has only made limited financial resources available for this purpose. Direct financial assistance for Natura 2000 has been provided through LIFE, the Community's

92   Under the Habitats Directive, Article 3(1).
93   Under Article 6(1).
94   European Commission, *Financing Natura 2000*, COM (2004)431, 5.
95   Beaufoy, G. (1999), *Natura 2000: Opportunities and Obstacles* (London: World Wildlife Fund Ltd) 15.

financial instrument for the environment.[96] Between 1992 and 2006 the European Community spent more than €1.44 billion co-funding 970 nature conservation projects in Member States.[97] However, the European Commission's estimates of the cost of managing the Natura 2000 network, as a whole, place this expenditure firmly in perspective.

The European Union has a vested interest in ensuring that the Natura 2000 network is properly managed, since these sites have a vital contribution to make in halting biodiversity loss. It, therefore, has a strong incentive to help Member States shoulder the financial costs involved. For its part the European Commission has suggested that additional Community funding for Natura 2000 should be provided through existing policies.[98] In particular, it has pointed to the CAP as having a pivotal role in this regard.[99] Two measures of central importance are the agri-environment scheme and the more recent Natura 2000 scheme.

Regulation 1257/99 extended the Community's less favoured areas scheme by providing Community funding for land management in areas in which farmers faced restrictions on the agricultural use of their land as a result of Community environmental protection legislation.[100] This gave Member States discretion to introduce national measures to compensate farmers for the costs they incurred and profits they lost as result of managing Natura 2000 sites. Under 1698/2005 this measure evolved into a stand alone Natura 2000 payment scheme.[101] The Regulation also removed a restriction that had previously required Member States to limit its availability to a maximum of 10 per cent of their land area.[102] In practice, however, Member States proved slow in adopting this measure. The European Commission reported that in 2001 only Germany, Italy and Spain had introduced national Natura 2000 payments, with 95 per cent of the farms that received payments being located in Germany.[103] The delays that have occurred in identifying and designating Natura 2000 sites may be partially responsible for this low uptake.[104] Given the importance of the Natura 2000 network to the Community's 2010 commitment, there may be

---

96   LIFE was established by Council Regulation 1973/92, [1992] OJ L206/1. It currently operates under Council Regulation 1655/2000, [2000] OJ L192/1, which was amended by Regulation 1682/2004 of the European Parliament and the Council, [2004] OJ L308/1.

97   Silva, J.P., et al. (2009), *Learning from LIFE: Nature Conservation Best Practices*, (Luxembourg: Office for Official Publications of the European Communities) 6, available at http://ec.europa.eu/environment/life/publications/lifepublications/lifefocus/documents/best_nat.pdf [accessed August 12, 2009].

98   European Commission, footnote 86 above, 8.

99   *Ibid.*

100   Council Regulation 1257/99, [1999] OJ L160/80, Article 13(b).

101   Council Regulation 1698/2005, [2005] OJ L277/1, Article 36(a) (iii) and 38.

102   *Ibid.*

103   European Commission Agriculture Directorate General (2004), *Biodiversity Action Plan for Agriculture: Implementation Report* (Brussels: European Commission) 20.

104   *Ibid.*

a case for making it compulsory, for all Member State rural development plans, to make provision for Natura 2000 payments.[105] Equally, the Community should reconsider the basis upon which scheme payments are made. Currently, Regulation 1698/2005 requires that the payments should compensate 'for costs incurred and income foregone' as a result of a Natura 2000 designation value of these sites in agricultural production terms.[106] It would be much more appropriate for the scheme to recognise that farmers should be paid for the environmental services that they, in turn, provide in managing these sites.[107]

The Natura 2000 payments scheme provides Member States with a measure that can help fund positive management on Natura 2000 sites. By contrast, the agri-environment scheme enables them to fund conservation management both on these sites and also across the remainder of the farm. As chapter 5 emphasised, the agri-environment scheme is not targeted specifically at nature conservation. It seeks, rather, to support farming methods that protect the environment as whole. Nevertheless, the scheme does provide a valuable opportunity for Member States to use Community finances to help fund the cost of managing Natura 2000 sites. One example of this is provided by Ireland's agri-environment scheme, the Rural Environment Protection Scheme (REPS). Farmers who decide to enter a REPS management agreement will be required to comply with additional management obligations if they have a Natura 2000 site on their land. In return, these farmers are eligible to receive higher payments for managing this part of their farm. Unfortunately, as conservation groups and academic commentators have pointed out, too many other Member States have failed to recognise the opportunity that the agri-environment scheme presents, to help fund such management.[108] Member States must, therefore, ensure that they design and target their own measures in a way that enables them to maximise the potential of the agri-environment scheme.

CAP's Support for Nature Conservation in the Wider Countryside

The development of compulsory cross-compliance measures has an important role to play in integrating concern for nature conservation into agricultural practice in the countryside as a whole. In particular, the requirement for Member States to develop codes of practice on maintaining land in good agricultural and environmental

---

105   Jack, B., The European Community and Biodiversity Loss: Missing the Target. [2006] 15 *Review of European Community and International Environmental Law* 304, at 314.

106   Council Regulation 1698/2005. [2005] OJ L277/1, Article 38(1).

107   *Ibid.*

108   See, for example, Birdlife International (1996), *Nature Conservation Benefits of Plans under the Agri-environment Regulation* (Sandy: Birdlife International); Lowe, P. and Baldock, D. Integration of Environmental Objectives into Agricultural Policy Making, in Brouwer, F. and Lowe, P. (eds) (2000), *CAP Regimes and the European Countryside* (Wallingford: CAB International) 45–48.

condition means that all farmers who receive direct income payments must now face this issue. In the previous section criticism was made of earlier attempts to require farmers to apply 'usual good farming practices', for example, within the less favoured areas scheme. As Community law established no framework for these measures, too few paid adequate attention to nature conservation issues. In contrast, the fact that Community law has now established a framework for national codes of practice should ensure that these issues are addressed. For example, this framework requires Member States to include measures that require farmers to provide a minimum level of maintenance and avoid deterioration of habitats upon their farms. Additionally, the fact that the Single Farm Payment is based upon historic production levels should also, indirectly, benefit nature conservation. Payments based upon historic production levels provide no incentive for farmers to continue to intensify those production levels.

However, familiar question marks also hang over the effectiveness of these measures. First, it should be remembered that not all farmers benefit from direct payments such as Single Farm Payments or Less Favoured Area Payments. Consequently, these cross-compliance measures will have no significance for some farmers. Secondly, it remains to be seen how effectively Member States will police and enforce the measures set out in their codes of agricultural and environmental good practice. Equally, the fact that Member States have been given discretion to retain production-based arable area payments and livestock payments means that some agricultural policies may continue to encourage some farmers to damage wildlife habitats in order to intensify their production. Finally, it should also be acknowledged that cross-compliance measures can only ever play a limited role in protecting nature conservation. It was noted in Chapter 3, that the European Community's environmental philosophy for the CAP is that farmers should observe basic environmental standards without compensation. Equally, it recognises that they should be paid for providing any further environmental services. In the context of nature conservation, this would indicate that cross-compliance can require farmers to alter agricultural production practices to avoid those that are most harmful to wildlife. However, it is more difficult to incorporate cross-compliance measures that require farmers to provide the positive management that will enhance the nature conservation value of the land. Such positive measures fall more easily into the category of environmental services that must be purchased through the Rural Development pillar of the CAP.

In Nature Conservation terms, the principal importance of the Community's Rural Development Policy lies in the encouragement that it can provide for positive nature conservation management. The two schemes that are particularly important in this regard are the less favoured area scheme and the environmentally sensitive area scheme.

As explained in Chapter 4, the less favoured areas scheme provides payments to support farmers based in agriculturally less favourable areas. The payments help to support nature conservation in these areas by helping to prevent farm abandonment and the loss of biodiversity that often follow such abandonment.

However, the scheme has previously encouraged environmental damage and, indeed, the very biodiversity loss that the European Community now seeks to halt. Additionally, as Fennell has pointed out, these environmental benefits have been largely incidental to the principally socio-economic objectives of the scheme.[109]

By amending the less favoured areas scheme to introduce the same cross-compliance conditions as apply under the Single Farm Payment, the European Community has sought to ensure that farmers incorporate basic measures into their practice to protect wildlife and, indeed, the broader environment. Equally, the move from headage to area-based payments has removed an important incentive for farmers to intensify their production and rewarded extensive farmers. However, it is the support that the scheme provides for extensive farmers that provides the principal environmental benefit of the scheme. The poorer agricultural conditions prevailing in some less favoured areas have prevented farmers from intensifying their production. Several reports have highlighted the fact that areas with traditional low intensity agriculture often support high nature conservation value farming systems.[110] Consequently, most areas of extensively farmed high nature value farmland are found in less favoured areas. In the case of the United Kingdom, some 89 per cent of our low intensity farmland is found within less favoured areas.[111] Across the European Community most areas of high nature value farmland are to be found in less favoured areas.[112]

As a result, these farmlands often retain semi-natural habitats and support a broad range of wildlife. Consequently, by supporting the continuation of extensive farming, the scheme provides important, if largely incidental, support for nature conservation.

In reality, the less favoured areas scheme provides inefficient support for extensive high nature value farmland. For example, some areas of high nature value farmland can be found outside less favoured areas. These farmlands, obviously, gain no benefit from the scheme. Additionally, as the European Environment Agency has

109   Fennell. R. (1997), *The Common Agricultural Policy: Continuity and Change* (Oxford: Clarendon Press) 345.

110   See Baldock, D. et al. (1993), *Nature Conservation and New Directions in the EC Common Agricultural Policy* (London: Institute for European Environmental Policy); Beaufoy. G. et al. (1994), *The Nature of Farming: Low Intensity Farming Systems in Nine European Countries* (London: Institute for European Environmental Policy) or Bignal. E. and McCracken, D., Ecological Resources of European Farmland, in Whitby, M. (ed.) (1996), *The European Environment and CAP Reform, Policies and Prospects for Conservation* (Wallingford: CAB International).

111   Wilson, J., The Extent and Distribution of Low Intensity Agricultural Land in Britain, in Joint Nature Conservation Committee (JNCC) (1991), *Birds and Pastoral Agriculture in Europe* (Peterborough: JNCC).

112   European Environment Agency (2004), *High Nature Value Farmland: Characteristics, Trends and Policy Challenges* (Copenhagen: European Environment Agency) 14.

pointed out,[113] no clear relationship exists between the amount of high nature value farmland located within a Member State and payment levels under the scheme in each Member State.[114] This is illustrated by a comparison of Tables 7.1 and 7.2:

Table 7.1   Share of High Nature Value Farmland Area in EU-15 (percentage of total utilised agricultural area)

> 30%	Greece, Portugal, Spain
20-30%	Ireland, Italy, Sweden, United Kingdom
10-20%	Austria, France
1-10%	Belgium, Denmark, Germany, Finland, Luxembourg, The Netherlands.

Source: European Environment Agency (2006), *Progress Towards Halting the Loss of Biodiversity by 2010* (Copenhagen: European Environment Agency) 38.

Table 7.2   Average Less Favoured Area Payments in the EU-15 (euros per hectare)

Finland	194
Austria	137
Luxembourg	131
Sweden	121
Italy	113
France	95
The Netherlands	91
Ireland	90
Portugal	75
EU-15 Average	**71**
Germany	66
Denmark	57
Greece	55
United Kingdom	43
Spain	22

Source: European Commission (2004), *The Agricultural Situation in the European Union, 2003 Report* (Luxembourg: Office for Official Publications of the European Communities) Table 3.6.2.2. © European Communities, 2004.

---

113   *Ibid.*
114   *Ibid.* 15.

A comparison of these tables suggests that Member States have not used the less favoured areas scheme as a means of supporting farming on high nature value farmland.[115] The tables also raise questions as to whether the scheme is being used effectively to achieve its primary goal of preventing farm abandonment. The European Environment Agency has pointed to Ireland, the south of Portugal, Northern Ireland, large parts of Italy and parts of Spain and France as being the areas with the highest risk of farm abandonment.[116] However, little correlation seems to exist between levels of payment and the level of risk.

It was noted, in Chapter 4, that the less favoured areas scheme seeks to compensate farmers for the production limitations they faced upon their farms. This reflects the production orientation of the CAP at the time the scheme was introduced. The CAP has now moved on. Today it seeks to promote increased quality over increased quantity. However, the philosophy of the less favoured areas scheme remains unchanged. When the scheme was introduced, the environmental value of many less favoured areas was, generally, not recognised. For example the European Commission's 1968 Mansholt plan would have forested most of these areas.[117] Now that this value has been recognised, much greater attention must be given to it.

The Court of Auditors has criticised the scope of the current less favoured areas scheme, arguing that not all of the 56 per cent of Community agricultural land that currently benefits from it is, actually, agriculturally less favoured.[118] It could be suggested that rather than simply re-examining the current less favoured areas, a more radical reform would replace the entire scheme with one that rewards farmers for providing environmental services. Viewed from the perspective of nature conservation, this would enable extensive farmers to be rewarded for the work that they do in supporting and maintaining wildlife habitats. It would, effectively, abolish the less favoured areas scheme and leave the environmentally sensitive areas scheme as the principal vehicle to support wildlife management. This, however, would only be a positive step if the additional financial resources it freed up were also transferred to the environmentally sensitive areas scheme.

In 1998, a study, carried out for the European Commission, recommended that all price supports and compensatory payments should be phased out and replaced by 'Environmental and Cultural Landscape Payments.'[119] Adopting this approach

---

115   See also European Environment Agency, *Ibid.* 15.

116   European Environment Agency, (2006), *Progress Towards Halting the Loss of Biodiversity by 2010* (Copenhagen: European Environment Agency) 35.

117   European Commission, *Memorandum on the Reform of Agriculture in the European Economic Community and Annexes* (1969) 1 Bull. E.C. 1.

118   Court of Auditors, *Special Report 4/2003, Concerning Rural Development: Support for Less Favoured Areas*, [2003] OJ C151/3.

119   Buckwell, A. et al. (1998), *Towards a Common Agricultural and Rural Policy for Europe*, European Economy and Studies No.5 (Luxembourg: Office for Official Publications of the European Communities).

would have meant that the only payments available to farmers under the CAP would have been those made in return for environmental management. Ultimately, these proposals were not followed by the European Community. One consequence of this is that the less favoured areas scheme remains in place and continues to provide little more than indirect support for nature conservation management. Instead, the role of providing more targeted support falls to the Environmentally Sensitive Areas Scheme.

As the European Commission itself has acknowledged, agri-environment measures 'play an essential role in the achievement of the Community's biodiversity objectives.'[120] The Commission also noted that 'the implementation of targeted agri-environmental measures on the whole EU's territory constitutes, now, the core of the Community's environmental strategy.'[121]

As Chapter 5 pointed out, more than 20 per cent of the Community's agricultural land is currently enrolled in an agri-environment scheme. There is no doubt that some agri-environment schemes have produced some very positive biodiversity benefits. The European Commission has highlighted a number of schemes that have done so.[122] For example, they note that in England populations of curl bunting increased by 82 per cent on land in which management agreements targeted the management of their habitats.[123] This clearly illustrates that the agri-environment scheme can make a very positive contribution towards halting biodiversity decline. Equally, it must also be recognised that the agri-environment scheme is not, solely, a nature conservation tool. Chapter 5 revealed that Member States have been able to design agri-environment schemes to address a range of environmental issues.

The European Commission's assessment of the national agri-environment measures in place in 2002, found that only 15 per cent of land enrolled was taking part in measures that specifically targeted biodiversity and landscape enhancement.[124] This, relatively low emphasis upon biodiversity and landscape enhancement, suggests that national agri-environment measures do not always promote the management practices that would be most effective in enhancing biodiversity. This would also seem to be shown by an examination of areas of high nature conservation value. As Table 7.1 illustrates, large areas of high nature value farmland are found in Spain and Portugal. However, the European Commission has assessed that less than 10 per cent of land enrolled in agri-environment schemes in these countries is participating in a scheme specifically

120   European Commission, *Biodiversity Plan for Agriculture*, COM (2001)162 final 16.

121   *Ibid.*

122   European Commission (2005), *Agri-Environment Measures: Overview on General Principles, Types of Measures and Application* (Brussels: European Commission Agriculture Directorate General) 16.

123   *Ibid.*

124   *Ibid.* 13.

targeting nature and landscape enhancement.[125] This contrasts with a figure of 50 per cent in the Netherlands, a Member State containing much less high nature value farmland. A similar picture emerges in relation to Member State expenditure on agri-environment schemes. As was the case previously in relation to less favoured areas, the European Environment Agency has, once again, reported that no clear relationship exists between levels of agri- environment expenditure per hectare and the share of high nature value farmland found in each Member State.[126] Equally worrying is the fact that some of the lowest levels of enrolment into the scheme have occurred in Member States with the largest areas of high nature value farmland. For example, by 2002, less than 10 per cent of farmland in Spain and less then 5 per cent of farmland in Greece had been enrolled.[127] In both cases the agri-environment scheme formed a relatively small part of each of these Member States' Rural Development policies, meaning that little funding was allocated to it. The European Commission has referred to the importance of targeted agri-environment measures. However, it would seem clear that, in fact, Member States have failed to develop sufficiently targeted agri-environment measures and have not always provided sufficient financial support for the scheme. Unless this situation changes it is unlikely that the scheme will fulfil its potential, to enable the Community to achieve its policy goal of halting biodiversity decline by 2010.

---

125   European Commission (2004), Working Document: *Biodiversity Action Plan for Agriculture – Implementation Report* (Brussels: European Commission Agriculture Directorate General) 10.

126   European Environment Agency, footnote 112 above, 12.

127   European Commission (2005), *Agri-Environment Measures: Overview on General Principles, Types of Measures and Application* (Brussels: European Commission Agriculture Directorate General) 7.

Chapter 8
Regulating Agricultural Pollution

Preventing Environmental Pollution

Previous chapters have outlined the environmental impacts associated with modern intensive agricultural methods and the steps taken to integrate environmental protection measures within agricultural policy. In contrast, this chapter examines a number of European Community environmental measures, outside the CAP, that require Member States to address pollution from agricultural sources. In each case, the European Community measures have been adopted in the form of directives. Chapter 7 highlighted the obligations imposed upon Member States by the Wild Birds and Habitats Directives. This chapter examines a number of other directives primarily concerned with water pollution and the contamination of land or soil. In the case of the 1991 Nitrates Directive and the 1986 Sewage Sludge Directive, the legislation specifically targets the agricultural industry. In other cases, the legislation applies more generally to all sectors of the economy.

Nitrate Pollution in European Waters

Agricultural intensification led to increased levels of animal manure being used on fields, as livestock numbers rose, and increased use of chemical fertilisers. By the early 1990s, chemical fertiliser usage in north-western Europe was reported to be the highest in the world.[1]

Livestock manures and chemical fertilisers provide nutrients, such as nitrogen and phosphorous, which are essential for plant growth. The process of intensification, however, often results in crops and grasslands receiving higher levels of nutrients than they can absorb. This is compounded by the fact that some farmers, such as intensive pig and poultry farmers may only have limited land on which to spread animal manure, leading to a situation in which this manure is, effectively, dumped as waste upon their land.[2] Such factors lead to soil nutrient surpluses. For example, it has been estimated that an average nitrogen surplus of

---

1   Agra Europe (1991), Agra Europe Special Report No.60 – *Agriculture and the Environment: How will the European Community Resolve the Conflict?* (London: Agra Europe Ltd) 9.

2   Dietz, F.J. Heinjnes, H., Nutrient Emissions From Agriculture, in Dietz, F.J., Vollebergh, H.R.J., Veries, J.L. (eds) (1993), *Environmental Incentives and the Common Market* (Dordrecht: Kluwer) 15.

60 kg per hectare existed across the European Community in 1997.[3] Some 95 per cent of this surplus is believed to have leeched into surface and groundwaters.[4] This, in turn, has an important influence upon nitrate levels found within rivers and groundwaters. These levels rose steadily throughout the 1970s and 1980s. The European Environment Agency, in a comparison of the periods 1977 to 1982 and 1988 to 1990, found increased nitrate levels in over two thirds of European rivers, with a median increase of 13 per cent.[5] More recently, nitrate levels are reported to have stabilised in the 1990s,[6] in the wake of reforms to the CAP. However, depending upon the catchment area, agriculture is still reported to be responsible for between 46 and 87 per cent of the nitrate loads found within European Community waterways.[7]

Nitrate enrichment in European waterways became a matter of concern to European Community policy makers as a result of the potential risks it posed to human health and the wider environment. Chapter 2 highlighted the health concerns associated with high concentrations of nitrate in drinking waters. In 1970, the World Health Organisation recommended a limit of 50 mg/l for nitrate levels within drinking waters.[8] This formed the basis for the European Community's own drinking water standards. In 1975, Directive 75/440, on the quality of surface waters intended for drinking water abstraction, required Member States to ensure that the nitrate content of these waters did not exceed the mandatory limit of 50 mg/l and also set a non-binding guidance target of less than 25 mg/l.[9] The Community's 1980 directive on the quality of water intended for human consumption subsequently adopted the same guidance target and mandatory limit.

Chapter 2 also highlighted the fact that nutrient enrichment also causes eutrophication in European waterways.[10] The nutrients often stimulate algal blooms on the surface of stagnant or slow moving waters. As noted in Chapter 2,

---

3   Crouzet, P., Nixon, S., Rees, Y. et al. (2000), *Nutrients in European Ecosystems* (Copenhagen: European Environment Agency) 28. See also Nixon, S., Trent, Z., Marcuello, C. et al. (2003), *Europe's Water: An Indicator Based Assessment* (Luxembourg: Office for Official Publications of the European Communities) 44.

4   European Environment Agency (2001), *Environmental Signals 2001* (Luxembourg: Office for Official Publications of the European Communities) 53.

5   Stanners, D. and Bourdeau, P. (eds) (1995), *Europe's Environment: The Dobris Assessment* (Copenhagen: European Environment Agency) 94.

6   European Environment Agency (2002), *Environmental Signals 2002: Benchmarking the Millennium* (Copenhagen: European Environment Agency) 92.

7   Crouzet, P., Leonard, J., Nixon S. et al. (2000), *Nutrients in European Eco-Systems* (Luxembourg: Office for Official Publications of the European Communities) 32.

8   World Health Organisation (1970), European *Standards for Drinking Water* 2nd ed. (Copenhagen: World Health Organisation, Regional Office for Europe) 28.

9   [1975] OJ L194/26.

10   See Crouzet, P., Nixon, S., Rees, Y. et al. (2000), *Nutrients in European Ecosystem* (Luxembourg: Office for Official Publications of the European Communities) 36.

these blooms can have a major impact upon the ecological diversity of those waters and upon human and animal health. Additionally, their unsightly appearance and, at times, unpleasant smell, can also have a detrimental economic impact, since people become less willing to make recreational use of affected waters.

Directive 91/676/EEC concerning the protection of waters against pollution caused by nitrates from agricultural sources ('the Nitrates Directive'), seeks to tackle agriculture's role in contributing to this water pollution.[11] The Directive provides Member States with a choice. On the one hand, they may identify individual waters affected by such pollution, or that could become affected in future.[12] On the other hand, they can opt to take protective measures throughout their territory.[13]

'Pollution' is defined as meaning:[14]

> the discharge, directly or indirectly, of nitrogen compounds from agricultural sources into the aquatic environment, the results of which are such as to cause hazards to human health, harm to living resources and to aquatic ecosystems, damage to amenities or interference with other legitimate uses of water.

Bodies of water will be regarded as being polluted if they have a nitrate content of more than 50 mg/l, or could do so if preventative measures are not taken.[15] Similarly, the fact that a body of water is eutrophic or at risk of becoming eutrophic in the future provides a further indicator of the pollution targeted by the Directive.[16] In each situation, the fact that particular bodies of water have been identified as being polluted creates an obligation for Member States to designate all lands draining into those waters as being 'Nitrate Vulnerable Zones'.[17] Alternatively, Member States who choose to apply protective measures throughout their territory will, effectively, be declaring their whole national territory to be one large Nitrate Vulnerable Zone.[18] Within each Nitrate Vulnerable Zone Member States must implement mandatory action programmes.[19] These programmes establish minimum standards for farming designed to reduce levels of nitrate pollution.

11　[1991] OJ L375/1, Article 1.

12　*Ibid.* Article 3(1) and Annex I.

13　*Ibid.* Article 3(5).

14　*Ibid.* Article 2(j).

15　*Ibid.* Article 3(1) and Annex I.

16　*Ibid.* Article 3(1) and Annex I.

17　*Ibid.* Article 3(2). Initial Nitrate Vulnerable Zones were to be designated by December 1993, with Member States being obliged to review these designations at least every four years.

18　*Ibid.* Article 3(5).

19　*Ibid.* Article 5(1). The initial action programmes were required to be in place by December 1999, for Nitrate Vulnerable Zones designated by December 1993. Where Nitrate Vulnerable Zones were designated subsequently, action programmes for these were to be in place within five years of their designation.

They require farmers to observe provisions regulating issues such as times of the year when land-spreading of manure and fertiliser is forbidden, the storage capacity required for livestock manure and the maximum amount of manure and fertiliser that can be applied to farmland.[20] Member States can seek permission to derogate from a maximum limit of 170 kg per hectare of nitrogen from animal manure established in the Directive. This has been achieved in Northern Ireland, where farmers can apply up to 250 kg per hectare as long as they observe specific conditions established by the European Commission.[21] The Directive also introduces an educational element for farmers located outside Nitrate Vulnerable Zones. Member States must establish voluntary codes of agricultural good practice setting out recommendations to reduce nitrate pollution from these farms.[22]

Elevated Nitrate Levels and the European Court of Justice

The 50 mg/l water nitrate limit is based upon drinking water standards recommended by the World Health Organisation and set out in European Community legislation. However, the Nitrates Directive applies to any bodies of water that have, or may in future have, nitrate levels that reach this threshold, not just those used as drinking water. The European Court of Justice made this abundantly clear in Case C-69/99 *European Commission v United Kingdom*, when the Court found the United Kingdom's implementation of the Directive to be inadequate, since it had only designated Nitrate Vulnerable Zones in relation to bodies of water used for drinking water abstraction that exceeded, or could in future exceed, the maximum nitrate content.[23]

One further issue that the Court has also settled is the extent to which it must be demonstrated that agriculture is the source of any elevated nitrate levels that may be found. This was established by the Court's decision in Case C-293/99 *R v Secretary of State for the Environment, ex parte Standley & Others*.[24] The case was brought by two farmers who challenged the methods, adopted in the United Kingdom, to identify catchment areas that had nitrate concentrations in excess of 50 mg/l. Their particular concern was that these catchment areas were being identified as being polluted under the Nitrates Directive if agricultural sources had made a 'significant contribution' to these nitrate levels. The farmers argued that this should only occur when the water's nitrate concentration came exclusively

---

20   *Ibid.* Annex III.

21   See the Nitrates Action Programme (Amendment) Regulations (Northern Ireland) 2008, SR 2008/196.

22   *Ibid.* Article 4. These codes of good practice were to be in place by December 1993.

23   [2000] ECR I-10979. The Court also found that no Nitrate Vulnerable Zones at all had been designated in Northern Ireland, even though at least one body of water had been identified as polluted.

24   [1999] ECR I-2603.

from agriculture. The European Court of Justice, in a preliminary reference from
the High Court, rejected this interpretation. It pointed out that this interpretation
would be contrary to the spirit and purpose of the Directive.[25] It confirmed that
Member States were required to identify waters as polluted under the Directive,
and then to designate Nitrate Vulnerable Zones, when the nitrate concentration
of those waters exceeded 50 mg/l, or were at risk of doing so, and agricultural
sources were considered to have made a 'significant contribution' to that nitrate
concentration.[26] The Court, however, declined to establish precise criteria by which
this could be judged.[27] National authorities will, therefore, have to determine
this issue on an individual basis. The Court also rejected claims that the Nitrates
Directive infringed the 'polluter pays' principle, the principle of proportionality
and fundamental property rights of those farming lands within designated Nitrate
Vulnerable Zones. In relation to the 'polluter pays' principle and the principle of
proportionality, the Court observed that the Directive required Member States to
take into account other sources of nitrate pollution when designing their action
plans and that it, therefore, fully complied with these principles.[28] In relation to the
protection of property rights, the Court adopted its traditional view that these were
not absolute rights.[29] They could, therefore, be restricted by measures adopted
to achieve one of the Community's objectives of general interest, provided that
those measures did not constitute a disproportionate and intolerable interference,
impairing the very substance of the rights themselves. In this case the Court
confirmed that the Nitrates Directive sought to protect public health, which it
found to be one of the Community's general objectives. Equally, as the Court had
already found, the provisions that it put in place also satisfied the requirements of
the principle of proportionality.

Eutrophication and the European Court of Justice

The Nitrates Directive defines eutrophication as being:[30]

> ...the enrichment of water by nitrogen compounds, causing an accelerated growth
> of algae and higher forms of plant life to produce an undesirable disturbance to

---

25   *Ibid.* para. 33.

26   *Ibid.* para. 40. See also Case C-258/00 *European Commission v France* [2002]
ECR I-5959 in which France responded to a Commission complaint against French law
requiring nitrogen inputs to be 'predominantly' of agricultural origin, by amending its law
to require that agriculture be a 'significant' source of these inputs.

27   *Ibid.* para. 38.

28   *Ibid.* paras. 46–52.

29   As the Court has previously confirmed in Case 44/79 *Hauer v Land Rheinland-
Pfalz* [1979] ECR 3747, para. 23, Case 265/87 *Hermann Schräder HS Kraftfutter GmbH
& Co. KG* [1989] ECR 2237 para. 15, Case C-280/93 *Germany v Council of the European
Union* [1994] ECR I-4973 para. 78.

30   Article 2(i).

the balance of organisms present in the water and to the quality of the water concerned.

This is broadly similar to the definition provided by the Urban Waste Water Treatment Directive, which regulates the impact that nutrient discharges from sewage treatment works can have upon eutrophication. There, eutrophication is defined as:[31]

> ...the enrichment of water nutrients especially compounds of nitrogen or phosphorous, causing an accelerated growth of algae and higher forms of plant life to produce an undesirable disturbance of the balance of organisms present in the water and to the quality of the water concerned.

The only practical difference between these definitions is that, whilst the Urban Waste Water Treatment Directive is concerned with nutrient enrichment by both nitrogen and phosphorous compounds, the Nitrates Directive is solely concerned with enrichment by nitrogen compounds.

Until recently, Member States have had little guidance on the interpretation of these definitions. As a result they often acted restrictively in identifying waterways that were either eutrophic or at danger of becoming eutrophic. However, two recent judgments from the European Court of Justice have established that Member States need to adopt wider, more precautionary measures. This arose from Case C-258/00, *European Commission v France* (hereinafter European *Commission v France (Nitrates)*), in which the Court considered the legal definition of eutrophication under the Nitrates Directive[32] and Case C-280/02, *European Commission v France* (hereinafter European *Commission v France (Waste Water)*), in which the Court considered the definition provided in the Urban Waste Water Treatment Directive.[33]

The two cases complement each other, each being concerned with different aspects of the legal definition of eutrophication. In European *Commission v France (Nitrates)*, the Court dealt with the question of when 'the enrichment of water by nitrogen compounds' should be considered to be 'causing an accelerated growth of algae and higher forms of plant life.' In contrast, European *Commission v France (Waste Water)* considered the issue of when this enrichment and accelerated growth could be said to 'produce an undesirable disturbance to the balance of the organisms present in the water and to the quality of the water concerned.' Taken together, these judgments have clarified the full extent of Member State responsibilities. In general terms, the Court noted, in *European Commission v France (Waste Water)*, that:[34]

---

31   Urban Waste Water Treatment Directive, Article 2(11).
32   Case C-258/00 *Commission v France* [2002] ECR I 5959.
33   Case C-280/02 *Commission v France* [2004] ECR I 8573.
34   *Commission v France (Waste Water)* para. 19 of the judgment.

there must be a cause and effect relationship between the enrichment by nutrients and the accelerated growth of algae and higher forms of plant life on the one hand and, on the other hand, between the accelerated growth and an undesirable disturbance of the balance of organisms present in the water and the quality of the water concerned.

The principal issue in *Commission v France (Nitrates)* was whether France had taken sufficient action to identify freshwater eutrophication. France pointed out that the objective of the Nitrates Directive was to control nitrate pollution from agricultural sources. Additionally, they argued that the definition of eutrophication required that accelerated algal and plant growth was being 'caused' by nitrogen compounds. It was their view that these provisions meant that Member States could only be required to act when nitrates from agricultural sources were the 'controlling' factor in the growth of the plants or algae associated with eutrophication. In other words, the growth could be controlled by tackling pollution from agriculture. They sought to argue that this excluded a number of fresh waters and coastal waters from the scope of the Directive. Firstly, in relation to freshwaters, France argued that it had been scientifically proven that enrichment by phosphorous, not nitrogen, was the cause of accelerated growth of algae and plants in estuaries and hard freshwaters. They claimed that scientific studies had shown that nitrogen was abundantly available in these waters, whilst it was phosphorous that was in limited supply. Consequently, it was, in their view, the availability of phosphorous that regulated eutrophication within these waters. Secondly, France accepted that nitrogen was the controlling factor in other waters, such as coastal waters. However, they pointed out that even in these waters some plants and algae were able to fulfil their nitrogen requirements from other sources such as drainage basins, bottom sediments and even the atmosphere. Where this was the case, they argued that applying the Nitrates Directive would have no effect, since agriculture was not the source of the nutrient that stimulated algal and plant growth.

In practice, France and the European Commission produced conflicting scientific evidence on the role played by nitrogen in freshwater eutrophication. But, without taking this evidence into account, the Court decided that the possibility of certain categories of water being excluded from the scope of the Nitrates Directive was 'incompatible with the logic and objectives of the Directive.'[35] The Court went further and adopted a very precautionary view of the role played by nitrate in freshwater eutrophication. It noted that 'notwithstanding the role that phosphorous *may* play in eutrophication, plant species whose growth is accelerated by nitrogen *may* appear in such waters.'[36] The Court went on to acknowledge that the Nitrates Directive gave Member States wide discretion in identifying polluted waters, but found that this discretion could not be used in a way that resulted in 'a large portion of nitrogen bearing waters falling outside the scope of the Directive.' This

---

35   *Commission v France* (Nitrate) para. 50.
36   *Ibid.* (emphasis added by the author).

would place the onus upon Member States to implement the Nitrates Directive on the assumption that nitrogen plays an active part in all freshwater eutrophication. This interpretation is supported by the Court's subsequent decision in *Commission v Ireland*.[37] The Court, here, responded to Ireland's argument that phosphates were the chief cause of eutrophication in inland freshwaters by simply observing that the Nitrates Directive 'required Member States to identify eutrophic freshwaters or freshwaters at risk of eutrophication in the near future if preventative action is not taken.'[38] Again, the Court appears to be operating under the assumption that nitrate has a role in causing freshwater eutrophication, so that Member States will have to address nitrate pollution should bodies of water be shown to be either eutrophic or at risk of becoming eutrophic.

In *Commission v France (Waste Waters)*, the European Court of Justice examined the issue of when a particular nutrient source should be considered to be 'significant' in the context of eutrophication. It held that the Urban Waste Water Treatment Directive would only be engaged when waste-water discharges made a significant contribution to eutrophication or the risk of eutrophication.[39] The Court has, therefore, signalled that the focus should be on a pollution source's potential impact upon eutrophication. The presence of nitrate from agricultural sources will, therefore, be important if that nitrate makes a significant contribution towards eutrophication or the risk of eutrophication.

This still leaves the question of what will be deemed to be a 'significant contribution' to eutrophication? Previously, in *Standley*, in relation to elevated nitrate levels, the Court stated that Community law could not set out precise criteria.[40] Effectively, the issue will turn on the facts of each case. However, the Court's judgment in *Commission v France (Urban Waste Waters)* has provided guidance on when one particular nutrient can be considered to be making a significant contribution to eutrophication, or the risk of eutrophication. In the case, the European Commission produced evidence that 40 per cent of the nitrogen compounds carried by the River Seine came from urban waste waters. France argued that the correct figure was only 28 per cent, but the Court found that even this lower figure still represented a significant contribution to eutrophication.[41] Additionally, the Court found that nitrogen inputs from urban waste waters, accounting for between 21 and 32 per cent of the nitrogen loads carried in other French rivers, were also making significant contributions to the eutrophication.[42] Perhaps more surprisingly, the Court also found that the fact that 9.8 per cent of

---

37   Case C-396/01 *Commission v Ireland* [2004] ECR I-2315.

38   *Ibid.* para. 43.

39   Case C-280/02 *European Commission v France (Waste Water)*, at para. 25 of the Court's judgment.

40   Case C-293/97 *R v Secretary of State for the Environment, ex parte Standley* [1999] ECR I-2603 at para. 38 of the Court's judgment.

41   Case C-280/02 *European Commission v France (Waste Water)*, para. 40.

42   *Ibid.* paras. 84–86.

the spring and summer nitrogen inputs into the waters of the Lorient Roadstead came from urban waste waters which also made a significant contribution towards the eutrophication being experienced in those waters.[43] In contrast, the European Commission itself had decided that the fact that 8.9 per cent of the nitrogen load carried in the waters of Saint Brieuc Bay came from urban waste waters did not amount to a significant contribution to eutrophication there. These figures serve to illustrate the impact of the Court's approach. In deciding that urban waste waters did make a significant contribution to eutrophication within the Lorient Roadstead, the Court also took account of the fact that this amounted to 374 tonnes of material. As the Court noted in *Standley*, each case must turn on its own facts. However, it seems clear from this case that a source will, definitely, be regarded as making a significant contribution to eutrophication where it accounts for more than 20 per cent of the total amount of a particular nutrient carried by a body of water. Equally, it would appear that this could also be the case where that source is responsible for approximately 10 per cent or more of that nutrient load, provided that this also amounts to a significant volume of nutrient.[44]

Additionally, in determining whether agriculture makes a significant contribution towards eutrophication or the risk of eutrophication, the Court has signalled that it will adopt a catchment-area based approach.[45] In other words, the obligation to designate Nitrate Vulnerable Zones will arise in relation to bodies of water which are not themselves eutrophic or at risk of becoming eutrophic, but which carry large amounts of nitrate from agricultural sources into other waters that are eutrophic or are at risk of becoming eutrophic. In *Commission v France (Nitrate)*, the Court, faced with conflicting evidence as to whether the waters of the Seine Bay were eutrophic, found that these waters contained high nitrate levels that contributed to eutrophication in the North Sea.[46] This was sufficient to require France to identify these waters as being polluted under the Nitrates Directive.[47] Consequently, Member States will have to adopt a broad catchment-area based approach when implementing the Nitrates Directive. The Directive will be engaged when nutrient in one part of the water catchment area contributes towards the eutrophication, or the risk of eutrophication, in another part of the catchment area.

In *Commission v France (Waste Waters)*, France refuted Commission allegations that it had failed to identify a number of eutrophic waters under the Urban Waste Water Treatment Directive. In this case France argued that the mere fact that nutrient enrichment was causing accelerated plant growth was not

---

43   *Ibid.* para. 77.

44   *Ibid.*

45   See also the previous case C-396/00 *European Commission v Italy* [2002] ECR I-10979 in which the Court first applied this approach in relation to the Urban Waste Water Treatment Directive.

46   Case C-258/00 *European Commission v France* (Nitrate) paras. 68–69.

47   *Ibid.* para. 70.

sufficient, in itself, to establish that particular waters were eutrophic. They pointed out that the legal definition of eutrophication also required it to be established that the algal and plant growth had produced 'an undesirable disturbance to the balance of the organism present in the water and to the quality of the water concerned'. The Court was, therefore, called upon to examine these aspects of the legal definition of eutrophication.

The Court divided the question of what was meant by 'an undesirable disturbance to the balance of the organisms present in the water' into two parts. Firstly, it examined what was meant by 'a disturbance to the balance of the organisms present in the water' and then, secondly, considered the circumstances that would lead to that disturbance being considered to be 'undesirable'. On the first issue, the Court noted that scientific reports, produced by the Commission, referred to the fact that, 'the equilibrium of an aquatic ecosystem is the result of complex interactions among the different species present and with the environment.'[48] The Court held that the proliferation of any one species of algae or plant would amount to a disturbance of the balance of the aquatic ecosystem and, therefore, to the balance of the organisms present in the water.[49] This led the Court to consider when such a proliferation should be considered to be undesirable. Here the Court adopted a purposive approach. It noted that the objective of the Urban Waste Water Treatment Directive was to protect the environment from the adverse effects of urban waste water discharges. Since the Directive had been adopted on the basis of Article 130S of the EC Treaty, the Court found that this objective should be interpreted in the light of the Community's own environmental policy objectives. As set out in Article 130R of the EC Treaty, now Article 174, these required that Community environmental policy should contribute toward preserving, protecting and improving the quality of the environment and protecting human health. The Court noted that this policy objective required the Community, 'to prevent, mitigate or eliminate the harmful effects of human activities on flora and fauna, soil, water, air, climate, landscape and sites of particular interest and on the health and quality of life of persons.' The Court, therefore, concluded that the objective pursued by the Urban Waste Water Treatment Directive went 'beyond the mere protection of aquatic ecosystems and attempts to conserve man, fauna, flora, soil, water, air and landscape from any significant harmful effects of the accelerated growth of algae and higher forms of plant resulting from discharges of urban waste water.'[50] The Court, therefore, found that the issue of whether any disturbance to the balance of organisms present in the water should be considered undesirable should be considered on the basis of whether there would be any significant harmful effects for flora, fauna, man, the soil, water, air or landscape.[51] It concluded that, 'species changes involving loss of ecosystem biodiversity, nuisances due to the

---

48   *Commission v France (Waste Water)*, para. 21.
49   *Ibid.*
50   *Ibid.* para. 16.
51   *Ibid.* para. 22.

proliferation of opportunistic macroalgae and severe outbreaks of toxic or harmful phytoplankton therefore constitute an undesirable disturbance of the balance of organisms present in the water.'[52] The Court also adopted a purposive approach in interpreting the requirement that there should be 'deterioration of the quality of the water concerned'. Once again, it found that this issue should not simply be considered on the basis of whether deterioration in water quality had had harmful effects for aquatic ecosystems. It should also be satisfied by 'deterioration of the colour, appearance, taste or odour of the water or any change which prevents or limits water uses such as tourism, fishing, fish farming, clamming and shellfish farming, abstraction of drinking water or cooling of industrial installations.'[53] Given the close parallels between the Nitrates Directive and the Urban Waste Water Treatment Directive, which were both adopted under Article 130S of the EC Treaty, the Court can also be expected to adopt an identical approach in identifying cases of eutrophication under the Nitrates Directive.

The Court, therefore, adopted a very wide interpretation in examining these final elements of the legal definition of eutrophication. The practical impact of this approach is evident from the manner in which the Court applied it to particular French waters. For example, the Seine estuary was found to be suffering from severe deoxygenisation over an area of almost 50 km with the result that it was unable to support migratory fish such as salmon and eel for six months of the year.[54] In the Court's view this 'clearly constituted an undesirable disturbance of the balance of the organisms in the water and to the quality of the water.'[55] The Seine Bay was shown to be experiencing a proliferation of algae-producing toxins liable to accumulate in shellfish. These were potentially dangerous to humans who consumed them.[56] For the Court, the presence of these toxins amounted to an undesirable disturbance of the balance of organisms present in the water. Equally, the fact that they resulted in periodic bans upon shellfish collection represented deterioration in the quality of the water.[57] The Court also took account of the potential impact of algal blooms upon the tourist industry. It noted that algal blooms in the Seine Bay resembled foam and damaged the coast's appeal to tourists.[58] In the Artois-Picardy basin, it found that algal blooms caused the water to become slimy, to change colour and to have a nauseating odour,[59] whilst it was satisfied that algal blooms in the Lorient roadstead were causing 'green tides' which rendered impossible normal tourist activities, such as bathing, fishing

---

52 *Ibid.* para. 23.
53 *Ibid.* para. 24.
54 *Ibid.* paras. 43–44.
55 *Ibid.* para. 45.
56 *Ibid.* para. 36.
57 *Ibid.* para. 39.
58 *Ibid.* para. 37.
59 *Ibid.* paras. 53, 56.

and hiking along the coast.[60] In each case, the algal blooms were held to be an undesirable disturbance of the balance of the organisms in the water, whilst the impact upon tourism established that there had been deterioration in water quality. Member States will have to adopt similarly broad approaches in identifying waters that are either eutrophic or at risk of becoming eutrophic in future.

Implementing the Nitrates Directive in the United Kingdom

The Nitrates Directive requires Member States to review their Nitrate Vulnerable Zone designations every four years, to take account of any changes in the status of particular bodies of water.[61] In both England and Northern Ireland recent reviews have resulted in substantial increases in the amount of land that has become subject to Nitrate Action plans. In 1996, only 8 per cent of agricultural land in England was designated as a Nitrate Vulnerable Zone, this increased to 55 per cent in 2002 and will rise to 70 per cent in 2009.[62] Farmers within new Nitrate Vulnerable Zones will be required to observe Nitrate Action plan requirements from 1 January 2010.[63] In Northern Ireland only three small Nitrate Vulnerable Zones were initially declared in 1999. This increased to 7 in 2001. More recently however, Nitrate Action Plans have applied across the whole territory of Northern Ireland.[64] Where the designations were initially limited to small areas with elevated water nitrate levels, the impact of Courts judgements on eutrophication have largely been responsible for the increased area of land that is now designated. In contrast, however, Wales and Scotland have adopted more limited approaches. Legislative proposals for Wales propose extending the area of Wales designated as Nitrate Vulnerable Zones from 3 per cent to 3.6 per cent.[65] The most recent review of Nitrate Vulnerable Zones in Scotland recommended that no change be made beyond the four Nitrate Vulnerable Zones already designated there.[66]

---

60   *Ibid.* para. 73.

61   Nitrates Directive 1991, Article 3(4), see generally, Cardwell, M., The Polluter Pays Principle in European Community Law and Its Impact on UK Farmers, [2006] 59 *Oklahoma Law Review* 93.

62   Under the Nitrate Pollution Prevention Regulation 2008, SI2008/2349.

63   *Ibid.*

64   See *The Protection of Water Against Agricultural Nitrate Pollution Regulations* (Northern Ireland) 2004, SR 2004/419.

65   See the Welsh Assembly Government (2007), *The Protection of Waters against Pollution from Agriculture: Consultation on the Implementation of the Nitrates Directive in Wales* (Cardiff: Welsh Assembly Government).

66   See Scottish Government (2005), *Nitrate Vulnerable Zones in Scotland: Review of Designations of Nitrate Vulnerable Zones* (Edinburgh: Scottish Government). The existing zones are the Aberdeenshire, Banff, Buchan & Moray NVZ, the Strathmore and Fife NVZ, the Lothian and Borders NVZ and the Lower Nithsdale NVZ.

The Water Framework Directive

Like the Nitrates Directive, the Water Framework Directive has an important role to play in relation to water pollution caused by agriculture. The 2000 Water Framework Directive set Member States the objective of ensuring that their surface waters achieve 'good surface water status' by December 2015, at the latest.[67] It also provides for Member States to ensure that their ground waters achieve 'good groundwater status' by that date. To achieve good surface water status, these bodies of water must be recognised as being of 'good ecological status' and also of 'good chemical status'.[68] Achieving good groundwater status will require that groundwaters have both a 'good quantitative status' and a 'good chemical status'. In each case, Annex V to the Directive provides extensive technical guidance on the particular standards that Member States must apply in assessing whether individual bodies of water have achieved these goals. Assessments of the chemical status of both surface and groundwaters will require national authorities to consider concentrations of pollutants carried in these waters. For surface waters, achieving good ecological status will require consideration of the composition and abundance of various life forms, whilst assessments of the quantitative status of groundwaters will require account to be taken of the impact of direct or indirect abstractions for agricultural, industrial or other purposes. Consequently, therefore, Member States will have to address any pollution forms, whether agricultural or otherwise, that may prevent bodies of water from meeting the requirements of good chemical or ecological status and any abstractions that threaten the quantitative status of groundwaters.

The Directive does include a limited number of exceptions that would allow Member States to fail to meet their objectives without being in breach of the Directive.[69] However, its general expectation is that the vast majority of Community waters must achieve good surface water quality or good groundwater quality status by 2015. The extent of the obligation placed upon Member States is also emphasised by the Water Framework's definitions of 'surface water' and 'ground water under the Directive. In terms of surface waters, the Directive applies to all inland waters, except groundwaters; and to transitional waters and coastal waters; whilst the requirement to achieve good chemical status also applies to territorial waters.[70] In relation to groundwaters, these are defined as including

---

67   Directive 2000/60/EC of the European Parliament and of the Council of 23 October 2000 establishing a framework for Community action in the field of water policy. OJ L327/1 2000, Article 4. Surface waters are defined, in Article 2(1) as meaning 'inland waters, except groundwater; transitional waters and coastal waters, except in the case of chemical status for which it will also include territorial waters'.

68   *Ibid.* Article 2(18) and Annex V.

69   *Ibid.* Article 4(5) to 4(7).

70   *Ibid.* Article 2(1).

all water below the surface of the ground in the saturation zone that is in direct contact with the ground or subsoil.[71]

The Water Framework Directive requires Member States to divide their territories into River Basin Districts, regulated by River Basin Authorities.[72] They were required to conduct an analysis of the characteristics of each River Basin District by 2004, to consider the characteristics of its bodies of water, the impact of human activity upon their water quality and to identify the various economic purposes for which the water was being used.[73] The River Basin Authorities must also establish a programme of measures for each of these River Basin Districts by 2012. At a basic level each programme must include the measures required in order to implement Community water protection directives, such as the Nitrates Directive, along with controls upon point source pollution and measures to prevent or control diffuse pollution.[74] Importantly, however, Member States are also required to take additional measures, if it appears that any body of water is unlikely to meet the target of achieving 'good surface water status' by December 2015.[75] This could well require them to adopt even stricter environmental standards than those set out in existing European Community legislation concerned with water quality or water pollution issues.[76] One such example of the impact of the Water Framework is evident in relation to the issue of phosphorous pollution from agricultural sources.

Agriculture is reported to be the source of approximately 50 per cent of the phosphorous carried in European rivers.[77] However, the European Community had previously sought to control phosphorous levels solely by regulating the content of sewage treatment discharges under the Urban Waste Water Treatment Directive. The Community's strategy was partially successful. The European Environment Agency has reported that phosphorous levels have generally been decreasing in European rivers and lakes.[78] However, they noted that this is not a universal trend. In particular, they identify Loughs Neagh and Erne in Northern Ireland as examples of bodies of water experiencing steadily increasing phosphate concentrations.[79] These increases have principally resulted from the fact that the issue of

---

71   *Ibid.* Article 2(2).

72   *Ibid.* Article 3(1).

73   *Ibid.* Article 5(1) and Annexes II and III. This analysis must also be updated no later than 2013 and also every 6 years thereafter.

74   *Ibid.* Article 11(3) Annex VI. It will also include other measures including those dealing with the quality of waters abstracted for drinking water and with the recovery of the costs of providing water services.

75   *Ibid.* Article 11(5)

76   *Ibid.*

77   ENDS Report 376, May 2006 at p.29.

78   European Environment Agency (2003), *Europe's Environment: The Third Assessment* (Copenhagen: European Environment Agency) 174–175.

79   *Ibid.* p.176. See also Stanners D., Bordeau, P. (1995), *Europe's Environment: the Dobříš Assessment* (Copenhagen: European Environment Agency) 336, which points to

phosphorous pollution from agriculture had not been addressed.[80] However, the Water Framework Directive has created an imperative for Member States to tackle such issues. If surface waters are to achieve the required 'good surface water status' by 2015 then one of the ecological criteria that they must satisfy is that they do not exhibit 'any accelerated growth of algae resulting in an undesirable disturbance to the balance of the organisms present in the water or to the quality of the water'.[81] In other words they will have to be free from eutrophication. Additionally, River Basin Authorities will have to establish nutrient limits for each body of water, at a level that will support this goal, and then ensure that these limits are not being exceeded.[82] As was noted above, the European Court of Justice found that Member States had an obligation to tackle nitrate pollution from agricultural sources when that pollution made 'a significant contribution towards eutrophication or the threat of eutrophication. Equally, the same principles would now seem to apply, under the Water Framework Directive, when phosphate enrichment, from agricultural sources, makes a significant contribution towards eutrophication or the threat of eutrophication. One consequence of this is that regulations have been adopted in Northern Ireland to limit farmers' ability to use chemical phosphorous fertilisers, in particular by requiring a soil analysis to show their necessity.[83]

Protecting Groundwaters

European Community Environmental Law has also taken specific action to protect groundwaters from deterioration and chemical pollution. This is due to the important role of groundwaters as a source of public water supplies in many parts of the European Community and to the influence that they have on the chemical status of the surface waters into which they flow. Member States were initially required to take action under Directive 80/68 on the protection of groundwaters.[84] The Water Framework Directive 2000 provides for Directive 80/68 to be repealed on 22 December 2013.[85] In the interim period, Directive 2006/118 on the protection

---

research by the National Agency of Environmental Protection in Denmark showing the need to control phosphorous pollution from agriculture in order to tackle eutrophication.

80   In the case of Lough Neagh see Foy, R.H. et al., Changing Perspectives on the Importance of Urban Phosphorous Inputs as the Cause of Nutrient Enrichment in Lough Neagh (2003), *The Science of the Total Environment* 87–99.

81   Water Framework Directive 2000, Annex V, table 1.2.

82   *Ibid.*

83   See the Phosphorous (Use in Agriculture) Regulations (Northern Ireland) 2006, SR 2006/488.

84   [1980] OJ L20/43. Implemented in Great Britain by the Groundwater Regulations 1998, SI 1998/2746 and in Northern Ireland by the Groundwater (Northern Ireland) Regulations 1998, SR1998/401. See also DEFRA (2001), *Groundwater Protection Code: Use and Disposal of Sheep Dip and Compounds* (London: DEFRA).

85   Directive 2000/60, [2000] OJ L327/1, Article 22.

of groundwaters has been adopted and will operate alongside the earlier directive until its repeal.[86]

The Water Framework Directive defines groundwaters as being 'all water which is below the surface of the ground in the saturation zone and in direct contact with the ground or subsoil'.[87] Under Directive 80/68 Member States are required to prevent the direct discharge of hazardous substances, identified in List I of the annex to the Directive, and to ensure that any disposal or tipping of these substances that might lead to indirect discharge was subject to prior investigation.[88] In the latter case it would then be prohibited unless the Member State's competent authority was satisfied that all technical precautions, necessary to prevent such indirect discharges, had been taken.[89] Of particular relevance to agriculture is the fact that the substances itemised in List I include organophosphorous compounds, which can be found in some sheep dip and pesticides. Consequently, this brought activities such as the disposal of waste pesticides and the disposal of sheep dip within the scope of the Directive. For its part, the European Court of Justice has confirmed that the Directive imposes an absolute prohibition on direct or indirect discharges of the substances identified by List I.[90] Additionally, Directive 80/68 also requires Member States to limit the introduction into groundwaters of substances listed in List II of the annex to the Directive, with the objective of avoiding water pollution.[91] Member States' competent authorities must conduct a prior investigation of all proposed direct discharges of such substances and of any proposed disposal or tipping of them that might lead to indirect discharges. [92] Authorisation can only be granted for such discharges, disposals or tipping if the competent authority is satisfied that all necessary technical precautions to prevent groundwater pollution have been taken.[93]

Directive 2006/118 adopts similar measures to prevent groundwater pollution. Additionally, it provides that in the period between it taking legal effect, 16 January 2009, and Directive 80/68 being repealed, 22 December 2013, all new authorisation procedures conducted under Directive 80/68 must also take into account the requirements of Directive 2006/118.[94] Under Directive 2006/118, Member States must take all measures necessary to prevent inputs of hazardous substances entering groundwaters.[95] In particular, they must ensure that such

---

86   [2006] OJ L372/19. At the time of writing the United Kingdom had not yet to implement this directive.

87   Directive 2000/60, Article 2(2).

88   Directive 80/68, Article 4(1).

89   *Ibid.*

90   See Case C-131/88 *European Commission v Germany* [1991] ECR I-825.

91   Directive 80/68, Article 5(1).

92   *Ibid.*

93   *Ibid.*

94   Directive 2006/118, Article 7.

95   Directive 2006/118, Article 6(1) (a).

measures are taken in respect of the families or groups of pollutants identified in points 1–6 of Annex VIII to the Water Framework Directive.[96] Once again, this list specifically identifies organophosphorous compounds. Additionally, Member States are also required to take all measures necessary to limit inputs to groundwaters of pollutants listed in the remaining paragraphs of Annex VIII and also any other non-hazardous pollutants that the Member State believes present an existing or potential risk of pollution.[97] Of particular relevance to agriculture is the fact that paragraph 11 of Annex VIII requires Member States to have regard to substances that contribute to eutrophication, in particular nitrates and phosphates. This provision, which was not included in List II of the annex to Directive 80/68, could influence controls upon the use of nitrate or phosphate based fertilisers in some regions.[98] Although these provisions refer to 'inputs' the Directive requires Member States to take into account diffuse pollution that has an impact upon the chemical status of groundwaters, where this is technically possible.[99] The Directive does include a number of exemptions, for example enabling Member States to allow inputs into groundwaters where they consider them to be of a quantity and concentration so small as to obviate any present or future danger of the deterioration of the quality of the groundwaters.[100]

Using Sewage Sludge as Fertiliser

Residual sludge from waste water treatment plants and septic tanks is rich in nutrients and can be used as an agricultural fertiliser. Practice varies widely across the European Community. In some areas more than half of all sewage sludge generated is applied to land, whilst in other areas little or no use is made of it.[101] Where it is used, environmental issues arise because sewage sludge can also contain toxic heavy metals, such as cadmium, copper, nickel, lead, zinc, mercury and chromium. The potential presence of these metals means that the use of sewage sludge in agriculture poses risks to human and animal health. To guard against these dangers, the Community adopted Directive 86/278 on the protection of the environment, and in particular of the soil, when sewage sludge is used in

---

96   *Ibid.*

97   *Ibid.* Article 6(1)(b).

98   List II in Directive 80/68 does contain nitrates, inorganic compounds of phosphorous and elemental phosphorous. However, it requires Member States to act to address the harmful effect of such substances on groundwaters. In practice eutrophication affects the surface waters into which groundwaters flow, rather than goundwaters themselves.

99   *Ibid.* Article 6(2).

100   *Ibid.* Article 6(3).

101   See European Commission, *Report on the Implementation of the Community Waste Legislation for the Period 2001–2003,* COM (2006)406 final 6.

agriculture.[102] The Directive sets out minimum measures, both in terms of the limits set for heavy metal concentrations in soil and sludge and the restrictions placed upon farming. Member States are then authorised to take more stringent measures.[103] In the United Kingdom the Directive has been implemented through the Sludge (Use in Agriculture) Regulations 1989,[104] which also applies to sewage sludge obtained from septic tanks.

Directive 86/278 is primarily concerned with the agricultural use of residual sewage sludge from sewage plants treating domestic or urban waste waters.[105] Member States are left to adopt their own measures concerning the use of sludge from septic tanks.[106] They must ensure that, unless injected or worked into the soil, sewage sludge receives treatment to reduce its harmful effects before being used in agriculture.[107] Member States are required to prohibit the use of sludge if soil concentrations of any of the heavy metals regulated by the Directive already exceed fixed limits.[108] They must also ensure that the agricultural use of sewage sludge will not itself cause the level of heavy metals found in the soil to exceed these limits. In this latter situation Member States have a choice. They can regulate the amount of sludge that can be applied to land and also ensure that the metallic content of that sludge does not exceed levels specified for particular heavy metals.[109] Alternatively, they can regulate the quantities of these metals that may be added to the soil over a period of time.[110] The United Kingdom chose to adopt the latter approach.[111]

At farm level, Member States must ensure that soil tests are carried out to establish both soil pH and the quantity of heavy metals already contained within

---

102   [1986] OJ L181/6.

103   Article 12. For an analysis of particular measures adopted   by individual Member States or regions see European Commission, Report from the Commission on the implementation of Community waste legislation for the period 1998–2000, COM(2003) 250 final/3, 87–92.

104   SI 1989/1263. For Northern Ireland see the Sludge (Use in Agriculture) Regulations (Northern Ireland) 1990, SR 1990/245.

105   Directive 86/278, Article 2(a)(i). Additionally, Article 3(2) provides for Member States to enact their own domestic legislation to regulate the use of residual sludge from septic tanks and similar installations for the treatment of sewage or the treatment of sewage from any other type of sewage plant.

106   *Ibid.* Article 3(2).

107   *Ibid.* Article 6.

108   *Ibid.* Article 5 and Annex 1A.

109   These levels are stipulated in Annex 1B.

110   *Ibid.* Article 5(2). Annex 1C.

111   Under the Sludge (Use in Agriculture) Regulations 1989, the average annual rate of addition of each regulated metal must not exceed a limit value established by the Regulations.

the soil.[112] The testing must, at least, be carried out before sewage sludge is applied, with Member States themselves left to determine the frequency of further testing.[113] Once sewage sludge has been applied to agricultural land, Member States must also regulate the use of that land.[114] The Directive requires them to prohibit farmers from using sewage sludge on grassland or forage crops within at least three weeks of use of that land for grazing or harvesting.[115] Additionally, sewage sludge cannot be applied to land used for the cultivation of fruit or vegetable crops normally eaten raw, which are grown in direct contact with the soil, within at least ten weeks of their harvest.[116] This does not include fruit trees.[117]

The European Commission has reported that average concentrations of heavy metals within sewage sludge used in agriculture within the European Community are, currently, well below the limit thresholds established within the Directive and are continuing to decline.[118] However, they also report that production of sewage sludge is rising, creating the possibility that greater quantities of sludge may be applied to land. For example, they predicted a 40 per cent increase in sewage sludge production between 1999 and 2005.[119] In the light of this increase the Community's thematic strategy for waste has announced the Commission's intention to bring forward proposals to revise Directive 86/278, with the intention of tightening the quality standards imposed under the Directive.[120]

Agricultural Waste

European Community law provides for the general control of waste under the terms of the 2006 Waste Framework Directive,[121] which consolidates reforms made to the earlier, now repealed, 1975 Waste Framework Directive.[122] Article 4 of the 2006 Directive requires Member States to take necessary action to:

---

112   The directive further requires testing to be conducted upon the sludge, to establish its pH and heavy metal content. See Article 9 and Annex IIA.

113   *Ibid.*

114   *Ibid.* Article 7.

115   *Ibid.* Article 7(a) This article requires Member States to fix the length of this period themselves, taking account of their own geographical and climatic conditions. However, the period must not be less than three weeks.

116   *Ibid.* Article 7(b) and 7(c).

117   *Ibid.*

118   European Commission, footnote 101 above, 6.

119   European Commission, Report from the Commission on the implementation of the Community waste legislation for the period 1995 to 1997, COM (1999) 752 final, 91.

120   European Commission, Thematic Strategy on waste and recycling, COM (2005)666.

121   Directive 2006/12, [2006] OJ L114/9.

122   Directive 75/442, [1975] OJ L194/39.

1. ...ensure that waste is recovered or disposed of without endangering human health and without using processes or methods which could harm the environment, and in particular:

(a) without risk to water, air or soil, or to plant or animals

(b) without causing a nuisance through noise or odours

(c) without adversely affecting the countryside or places of special interest

2. ...prohibit the abandonment, dumping or uncontrolled disposal of waste

For these purposes the Directive defines waste as being any substance or object that the holder discards or intends to discard.[123] This definition mirrors that which had previously applied under the 1975 Waste Framework Directive and has been the subject of extensive analysis by the European Court of Justice.[124] This case law has examined whether, on the facts of each particular case, a substance or object can be said to have been discarded. In general, the Court has adopted a wide interpretation, in line with the environmental protection objectives espoused by the Directive. In relation to agricultural practice, the Court has found that spreading slurry on fields used in agriculture was not a disposal of waste, 'where it is used as a soil fertiliser as part of a lawful practice of spreading on clearly identified parcels of land and if its storage is limited to the needs of those spreading operations.'[125] The fact that the evidence adduced in the particular case suggested that slurry was being applied without regard to the nutrient requirements of particular plants, also did not induce the Court to find that this amounted to a disposal of waste.[126] Similarly, the fact that the amount of slurry being spread exceeded limits introduced under an action plan adopted under the Nitrates Directive was viewed by the Court as being a breach of that Directive, rather than as evidence of a disposal of slurry as waste.[127]

The Waste Framework Directive specifically excludes certain types of agricultural waste from the scope of the Directive, this relates to 'animal carcasses,

---

123   Directive 2006/12, Article 1(a). The definition also refers to this substance or object falling into one of sixteen categories of waste listed in Annex I to the Directive. However, since the sixteenth category is a catch all provision, referring to 'any materials, substances or products which are not contained in the above categories' this is of little assistance.

124   See, for example, Case C-304/94 Euro Tombesi [1997] ECR I-3561; Case C-126/96 *Inter-Environnement Wallonie v Regione Wallone*,[1997] ECR I-7411, Case C-418/97 *ARCO Chemie Nederland Ltd* [2000] ECR I-4475, Case C-416/00 *Palin Granit Oy* [2002] ECR I-3533.

125   Case C-416/02 *European Commission v Spain* [2005] ECR I-7487, para. 89.

126   *Ibid.* para. 94, though interestingly the Advocate General's opinion had viewed this as evidence that the slurry was actually being discarded as waste. See para. 41 of the Advocate General's opinion.

127   *Ibid.* paras 96–97.

faecal matter and other natural, non-dangerous substances used in farming.'[128] However, this exclusion only applies to the extent that these materials are covered by other European Community legislation.[129] The principal European Community legislation on these issues is Regulation 1774/2002 laying down health rules concerning animal by products not intended for human consumption.[130] The Regulation defines animal by products as being 'entire bodies or parts of animals or products of animal origin not intended for human consumption, including ova, embryos and semen.'[131] The Regulation established rules for the collection, transport, storage, handling, processing, use or disposal of animal by products.[132] For example, fallen animals should be collected, transported and incinerated by operators who comply with the terms of the Regulation. As a derogation, Member States can allow animal by products to be burnt or buried on their home farms within 'remote areas'. However, Member States must first notify the European Commission as to how they intend to operate this derogation and also advise the Commission of the areas that they intend to classify as being remote, along with their rationale for so doing.[133]

Agricultural waste was originally exempt from the scope of the United Kingdom's waste management laws.[134] This, however, resulted in a successful European Commission enforcement action against the United Kingdom.[135] Subsequently, each United Kingdom region has incorporated agricultural waste within the general scope of its waste management regime.[136] However, these measures enable farmers to register with the relevant regulator, the Environment

---

128   Directive 2006/12, Article 2(b).

129   *Ibid.*

130   Council Regulation 1774/2002 [2002] OJ L273/1 as subsequently amended.

131   *Ibid.* Article 2(1) (a).

132   For the implementation of these rules in the United Kingdom see the Animal By-Products Regulations 2005, SI 2005/2347; the Animal By-Products (Wales) Regulations 2006, SI 2006/1293 (W.127), the Animal By-Products Regulations (Northern Ireland) 2003, SR 2003/495 and the Animal By-Products (Scotland) Regulations 2003, SSI 2003/411.

133   *Ibid.* Article 24(1). In the case of animals suspected of being infected by TSE or killed as part of a TSE eradication programme, it is an additional requirement, under Article 24(5), that the Member State's competent authority should supervise the burning or burial and that it is satisfied that the methods adopted have precluded all risk of the transmission of TSE.

134   As set out in Great Britain in the Environmental Protection Act 1990 and in Northern Ireland in the Waste and Contaminated Land (Northern Ireland) Order as amended by the Waste (Amendment) (Northern Ireland) Order 2007.

135   See Case C-62/03 *European Commission v The United Kingdom, Times Law Reports* 6 January 2005.

136   Today, see the Environmental Permitting (England and Wales) Regulation 2007, SI 2007/3538, the Waste (Scotland) Regulations, SSI 2005/22 and the Waste Management Licensing Regulations (Northern Ireland) 2003, SR 2003/493 as amended by SR 2006/280, SR 2006/489 and SR 2008/18.

Agency in England and Wales, the Scottish Environmental Protection Agency or the Northern Ireland Environment Agency, to gain exemptions enabling them to continue carrying out particular activities that would, otherwise, come within waste management law. Providing they register for the particular exemption, abide by the conditions relevant to it and operate the exemption in a manner that will not cause environmental pollution, harm to human health or adversely affect the countryside or places of special interest (such as Sites of Special Scientific Interest, Areas of Outstanding Natural Beauty, Natura 2000 sites or designations based upon International Law, for example Ramsar sites) they will be exempt from those waste management laws.

In addition to requirements imposed by the Waste Framework Directive in relation to general waste, Directive 91/689 on hazardous waste also creates additional obligations upon Member States to ensure that hazardous wastes are disposed of safely.[137] The Directive lists the properties that will lead to waste being considered hazardous. Additionally, the Council has also drawn up a list of particular types of hazardous waste.[138] This includes a range of material relevant to agricultural practice, such as sheep dip, concentrated pesticides, tractor batteries or machinery oils and veterinary products such as syringes or dressings containing animal body fluids.[139] In England, hazardous wastes are defined as being any waste listed as hazardous in the List of Wastes (England) Regulations 2005[140] or classified as such by the Secretary of State or other regulations.

Member States must ensure that hazardous wastes are not mixed with non-hazardous materials or that different categories of hazardous waste are not mixed together.[141] They must also ensure that hazardous wastes are properly packaged and labelled and that producers and transporters maintain detailed records of all consignments.[142]

---

137   [1991] OJ L377/20. The Directive has been implemented in the United Kingdom by the Hazardous Waste (England) Regulations 2005, SI 2005/894, the Hazardous Waste (Wales) Regulations 2005, SI 2005/1806 (W.138), the Special Waste (Scotland) Regulations 1996 and the Hazardous Waste Regulations (Northern Ireland) 2005 SR 2005/300 as amended.

138   See Council Decision 94/904, [1994] OJ L356/14.

139   In England, for example, the Hazardous Waste (England) Regulations 2005 define hazardous waste as being any waste listed as hazardous in the List of Wastes (England) Regulations 2005 (SI 2005/895) or classified as such by the Secretary of State or other regulations.

140   SI2005/895.

141   Directive 91/689, [1991] OJ L377, Article 2(2).

142   *Ibid.* Article 5.

Environmental Liability

The European Community has had a long-standing interest in establishing a Community-wide system of civil liability for environmental damage.[143] This was viewed as a mechanism through which polluters could be brought face to face with the costs of remedying the environmental damage that they were causing and to deal with well-recognised deficiencies within existing measures, such as the law of Tort in the United Kingdom.[144] In 2004, the European Community finally adopted the Environmental Liability Directive.[145] Several public consultations have been conducted on the implementation of the Directive in the United Kingdom, but at the time of writing implementing legislation had yet not been adopted.

The Environmental Liability Directive has been criticised for being much less ambitious than previous legislative proposals, which had sought to create more comprehensive environmental liability provisions across the Member States.[146] It essentially requires Member States to establish administrative measures to require particular polluters to pay the cost of either acting to prevent environmental damage from happening or of restoring the environment when damage does occur. This is clearly an application of the 'polluter pays' principle.[147] It also serves to encourage other operators to minimise any environmental risks associated with their businesses.[148] The Directive establishes minimum standards for the prevention and remediation of environmental damage and authorises Member States to maintain or adopt more stringent measures.[149]

The Directive applies to 'operators', a term defined as being any person, whether legal, private or public, who operates or controls an occupational activity.[150] This is clearly wide-enough definition to include farmers. The environmental liability of operators is determined by a two-tier system. Firstly, the Directive

---

143   See European Commission, *Communication from the Commission to the Council and the Parliament on Remedying Environmental Damage*, COM (93) 47 final and European Commission, *White Paper on Environmental Liability*, COM (2000) 66 final.

144   See Lee, M. Tort, regulation and environmental liability, [2002] 22 Legal Studies 33–52 or Wilde, M. EC Commission's White Paper on Environmental Liability: Issues and Implications, [2001] 13 *Journal of Environmental Law* 21–37.

145   Directive 2004/35/EC of the European Parliament and the Council on environmental liability with regard to the prevention and remedying of environmental damage, [2004] OJ L143/56.

146   See, for example Lee, M., footnote 144 above, at 34 and Bell, S., McGillivray, D. (2008), *Environmental Law*, 8th ed. (Oxford: Oxford University Press) 363.

147   As stated in paras 2, 18 of the preamble to the Directive.

148   *Ibid.*

149   Article 16.

150   *Ibid.* Article 2(6), additionally the definition provides for the term to include persons to whom decisive economic power over the functioning of that activity, where this is provided for in national legislation. This could include the holder of a permit or authorisation to carry out an activity or the person registering or notifying that activity.

imposes a system of strict liability where environmental damage or the imminent threat of such damage, results from particular activities listed in Annex III to the Directive.[151] Here, therefore, the mere occurrence of the damage or threat of damage will be sufficient to give rise to liability. Secondly, when environmental damage or the imminent threat of such damage stems from any other activities, environmental liability will only arise if it can also be established that the operator has been negligent or otherwise at fault.[152]

Annex III lists 12 activities, regulated by European Community environmental measures, to which strict liability is to apply. A number of these are directly relevant to farming. These include:

- The operation of installations that are subject to permit requirements under Directive 96/61 concerning integrated pollution prevention and control.[153]
- Conducting waste-management operations that are subject to permit or registration requirements under Directive 75/442 on waste[154] or Directive 91/689 on hazardous waste.[155] Member States have discretion to decide that this will not include the spreading of sewage sludge for agricultural use where that sludge meets the standards set under the Sewage Sludge Directive.
- Discharges into inland waters requiring prior authorisation under Directive 76/464 on pollution caused by dangerous substances.[156]
- Discharges of substances into groundwater that require prior authorisation under Directive 80/68 on groundwater pollution.[157]
- Discharges or injections of pollutants into surface or groundwaters that require a permit or authorisation under Directive 2000/60 establishing a framework for Community Water Policy.[158]
- Water abstraction or impoundment that requires prior authorisation under Directive 2000/60.
- The manufacture, use, storage, processing, filling, release into the environment and onsite transportion of plant-protection products (as defined in Article 2(1) of Directive 91/414 concerning the placing of plant-protection products on the market)[159]or biocidal [pesticide] products (as defined in Article 2(1) (a) of Directive 98/8 concerning the placing of biocidal products on the market).[160]

---

151 Article 3(1)(a).
152 Article 3(1)(b).
153 [1999] OJ L257/26.
154 [1975] OJ L194/39.
155 [1991] OJ L377/39.
156 [1976] OJ L129/23.
157 [2000] OJ 332/91.
158 [2000] OJ L327/1.
159 [1991] OJ L230/1.
160 [1998] OJ L123/1.

- Deliberate release into the environment, transportation and placing on the market of genetically modified organisms as defined in Directive 2001/18 on the deliberate release of GMOs into the environment.[161]

There are a number of exceptions to the strict liability that arise when environmental damage results from these activities. These arise when the environmental damage or the imminent threat of that damage results from an act of armed conflict, hostilities, civil war or insurrection or from a natural phenomenon of exceptional, inevitable and irresistible character.[162] Perhaps more importantly, Member States can decide not to impose liability upon operators where environmental damage has occurred and the operators can demonstrate that they were not at fault or negligent and that the damage was caused either by emissions that complied with the national laws implementing the relevant European Community measures or by a product that, on the basis of scientific knowledge at the time of the emission, was not considered likely to cause environmental damage.[163] Additionally, the Directive provides that environmental liability can only arise in relation to diffuse pollution if a causal link can be established between the damage, or threat of damage, and the activities of individual operators.[164]

In practice, however, the concept of environmental damage under the Directive is very limited in scope. Essentially, the types of environmental damage covered by the Directive are restricted to:[165]

a) Damage to species and natural habitats, protected under the Wild Birds Directive and the Habitats Directive, that is likely to have significant adverse effects in their reaching or maintaining a favourable status, unless authorised under the terms of these Directives

b) Damage to species and natural habitats protected under national nature conservation laws, such as SSSIs in Great Britain

c) Damage to bodies of water protected under the Water Framework Directive and likely to have significantly adverse effects upon their ability to achieve the environmental objectives set by that Directive

d) Damage to land through contamination, creating a significant risk to human health

In each case 'damage' is defined as, 'a measurable adverse change in a natural resource, or measurable impairment of a natural resource service, which may occur directly or indirectly.'[166] In addition, the Directive only applies within a

---

161   [2001] OJ L106/1.
162   Article 4(1).
163   Article 8(4)
164   Article 4(5).
165   *Ibid.* Article 2(1).
166   *Ibid.* Article 2(2).

limited time frame. In this sense, the Directive makes it clear that it does not apply to damage caused by an emission, event or incident that took place before 30 April 2007, the date by which Member States were required to have implemented the Directive.[167] Similarly, it does not apply where the environmental damage was caused by an emission, event or incident that took place after that date but derived from a specific activity that took place and finished before that date.[168] Finally, the Directive also makes it clear that liability will not attach to damage when more than 30 years have passed since the emission, event or incident occurred which resulted in that damage.[169]

Member States must designate competent national bodies tasked with identifying operators who cause environmental damage, or create an imminent threat of damage and deciding what action should be taken.[170] Other parties also have the right to notify this body of environmental damage that has occurred or of an imminent threat of damage occurring. This can be done by any natural or legal person who is affected or likely to be affected by this environmental damage or who has a 'sufficient interest' in environmental decision making concerning it.[171] The question of what amounts to a sufficient interest must be decided by each Member State.[172] However, non-governmental organisations promoting environmental protection and meeting any requirements imposed under national law are considered to possess sufficient interest.[173] The notifier must also be informed as to whether the authority decided to take action and of its reasons for deciding to act or to refusing to act.[174] This decision can then be challenged before a court or other independent and impartial public body.[175] These provisions provide individuals and environmental NGOs with an indirect role in the enforcement procedure. However, the role accorded to environmental NGOs is really much weaker than that which had previously been imagined. These would have enabled NGOs to bring proceedings in any situation in which a Member State's own competent authority had failed to do so.[176] Indeed, the relative weakness of this aspect of the current measures is emphasised by the fact that the Directive allows Member States to decide not to grant citizens or environmental NGOs a right to make complaints about situations in which they believe there is an imminent threat of environmental damage.[177]

---

167   *Ibid.* Article 17.
168   *Ibid.*
169   *Ibid.*
170   *Ibid.* Article 11.
171   *Ibid.* Article 12.
172   *Ibid.*
173   *Ibid.*
174   Article 12(4).
175   Article 13.
176   See European Commission, *White Paper on Environmental Liability,* footnote 143 above, 22. See also Wilde, M., footnote 144 above, 31.
177   Article 12(5).

Operators are required to act immediately to take necessary measures to prevent an imminent threat of environmental damage and, where damage has occurred, to manage any environmental damage that has occurred so as to limit or prevent further environmental damage and adverse impact on human health or natural resources.[178] In these situations, in addition to requiring operators to provide information, national authorities may also require operators to follow their instructions on appropriate measures that should be taken or to allow the national authority itself to take the necessary measures.[179] Annex II to the Directive also sets out a common framework to be used in deciding what the most appropriate measures are to ensure that environmental damage is remedied. The Annex distinguishes between primary remediation, complementary remediation and compensatory remediation. The Directive gives priority to primary remediation, under which environmental resources are restored to the condition they would have been in had the environmental damage not occurred. Where this is not possible then complementary remediation will seek to provide a similar level of environmental resource as would have existed had primary remediation been possible. This may well be at a different site. Finally, compensatory remediation seeks to achieve additional improvements, to protected habitats and species or bodies of water, to compensate for the interim loss of environmental resources whilst they are recovering.

Under the Directive, operators must bear the costs of all preventative and remedial measures that are taken and national authorities must act to recover those costs.[180]

This principle of cost recovery will not apply when the operator can prove either, that the damage or imminent risk of damage was caused by a third party and occurred despite there being appropriate safety measures in place, or that it resulted from steps that they had been instructed to take by a public authority.[181] National authorities will always be required to act to recover the cost of preventative and remedial measures unless the cost of doing so would exceed the monies that would be recovered or the operator cannot be identified.[182] In contrast, where an operator can establish that one of the above exceptions applies and that the cost of preventative or remedial action should, therefore, not be recovered from them, national authorities will then be under an obligation to help the operator to recover any preventative or remedial costs that it incurred from the third party or public authority concerned.[183]

---

178   Articles 5(1) and 6(1).
179   Articles 5(3) and 6(1).
180   Article 8(1) and 8(2).
181   Article 8(3).
182   Article 8(2).
183   Article 8(3).

Environmental Impact Assessment

The European Community's Directives on Environmental Impact Assessment also have implications for agriculture, though their impact here is more limited than in other economic sectors. Directive 85/337 on the assessment of the effects of certain public and private projects on the environment first introduced the concept into European Community law.[184] This Directive was, subsequently, amended by Directive 97/11/EC.[185] Essentially, the Directives put in place a mechanism to gather information about the environmental impact of proposed developments. Member States are placed under a duty to ensure that projects, likely to have significant effects on the environment by virtue of factors such as their nature, size or location, are made subject to a requirement for development consent and to a prior assessment of their likely environmental effects.[186] The environmental assessment is based upon an environmental statement provided by the person seeking development consent. The environmental statement should identify the direct and indirect impact that the project is likely to have upon humans, fauna, flora, soil, water, air, the landscape, material assets and cultural heritage.[187] The environmental statement must, at least, provide information on the following:[188]

- The site, design and size of the proposed project
- The measures that will be taken to avoid, reduce or remedy any significant adverse impacts upon the environment
- Data which identifies and assesses the principal environmental effects of the proposed project
- A description of the main alternatives considered by the developer and of the main reasons for their choice of preferred methods
- A non-technical summary of all the above information.

Ultimately, the procedure is designed to highlight the potential environmental impacts to those tasked with deciding whether to authorise the project. As such, it will be one material consideration that these decision makers will take into account when deciding whether to permit the development. The Directive also provides for public participation in this decision-making process, The public must be informed that a request has been made for development consent that is subject to an environmental impact assessment and the information supplied by the developer must be available to the public within a reasonable time to enable those concerned to express their opinions before any development consent is granted[189]

---

184   [1985] OJ L175/1.
185   [1997] OJ L73.
186   Directive 85/337, Article 2(1).
187   *Ibid.* Article 3.
188   *Ibid.* Article 5(3).
189   *Ibid.* Article 6(2).

Similarly, authorities with specific environmental responsibilities, who are likely to be concerned by the proposed development, must also be given the opportunity to express their opinion upon the proposal.[190] The decision-making body will have to make their decision public, highlighting any conditions attached to the decision, the main reasons and considerations upon which the decision was based and a description of the principal measures that will be taken to avoid, reduce or offset any major environmental effects from the project.[191] Courts in the United Kingdom have previously been quite dismissive of these public consultation requirements, upholding decisions to grant planning permission despite the absence of an environmental impact assessment, on the basis that all the necessary environmental information had been available to the planning authority, despite the absence of an environmental statement or public consultation.[192] This, however, altered with the House of Lords decision in *Berkley v Secretary of State for the Environment, Transport and the Regions*, in which it emphasised the need to ensure full public consultation.[193]

The Directives distinguish between projects that must be subject to an environmental impact assessment and projects in which there is discretion to require this assessment to be made. Annex I to the 1985 Directive, as amended by Directive 97/11/EC, contains a list of major projects that are considered likely to always have a significant environmental impact and for which an environmental impact assessment will be compulsory, whilst Annex II identifies other projects that may well have significant environmental effects. National authorities are then called upon when individual proposed developments, that come within the Annex II categories, are likely to have a significant impact upon the environment and should, therefore, be the subject of an environmental impact assessment. In relation to agriculture; the following projects have been included within Annex I and Annex II.

Annex I   Environmental Impact Assessment Compulsory

Installations for the intensive rearing of poultry or pigs with more than:
a) 85,000 places for broilers, 60,000 places for hens
b) 3,000 places for production pigs (over 30 kg)
c) or 900 places for sows.

---

190 *Ibid.* Article 6(1).
191 *Ibid.* Article 9.
192 See, for example, *R v Poole Borough Council*, ex parte Beebee [1991] JPL 643.
193 [2000] 3 All ER 897.

Annex II   Environmental Impact Assessment Discretionary

a) Projects for the restructuring of rural-land holdings
b) Projects for the use of uncultivated land or semi-natural areas for intensive agricultural purposes
c) Water-management projects for agriculture, including irrigation and land-drainage projects
d) Initial aforestation and deforestation for the purposes of conversion to another type of land use
e) Intensive livestock installations (those not covered in Annex I)
f) Intensive fish farming
g) Reclamation of land from the sea

The exercise of discretion, in deciding whether an environmental impact assessment will be required for an Annex II project, will require the competent authority to take a 'screening' decision as to whether they believe that it will be likely to have significant environmental effects. Previously, Directive 85/337 allowed Member States to provide for this decision to be based upon nationally established criteria and/or thresholds. However, the European Court of Justice, subsequently made clear that in doing so Member States have a fundamental obligation to ensure that projects likely to have significant environmental effects were subject to an assessment. Therefore, for example, it was wrong to set such thresholds at levels that would, in practice, exclude all potential Annex II developments from the scope of the Directive.[194] Directive 97/11/EC then introduced measures designed to limit Member State discretion concerning Annex II projects. It set out a number of selection criteria that are to be applied when deciding whether a particular Annex II project should be required to undergo an environmental impact assessment. These criteria fall into three categories:

1. *Characteristics of the development:* This includes considerations such as its size, its use of natural resources, its waste production and pollution potential.
2. *Location of the development:* This includes considerations of the environmental sensitivity of the area that will be affected by the development.
3. *Characteristics of the potential impact:* The potential environmental impact of the proposed development considered against factors such as the extent of that impact, its transboundary nature, its magnitude and complexity, the degree of probability that it will occur and its likely duration, frequency and reversibility.

---

194   See C-75/95 Kraaijeveld [1996] ECR I-5403 and Case C-392/96 *European Commission v Ireland* [1999] ECR I-5901.

These criteria are now applied in conjunction with nationally-established thresholds. Where the characteristics of the proposed development exceed these thresholds then an environmental impact assessment will, probably, be required. However, where those characteristics do not exceed the thresholds then the above factors will be considered. Where they indicate that the proposed development is likely to have significant environmental effects then an environmental impact assessment will be required.

In the United Kingdom, the Environmental Impact Assessment Directives have been implemented predominantly through development control law and policy.[195] This requires compulsory environmental impact assessments for projects falling within Annex I of the Environmental Impact Assessment Directives and sets thresholds for Annex II projects. For projects concerning agriculture these are shown in Table 8.1.

Table 8.1   Environmental Impact Assessment: Agricultural Project Thresholds

Project	Threshold
1. The use of uncultivated land or semi-natural land for intensive agricultural use	Area of proposed development exceeds 0.5 hectares
2. Water-management projects for agriculture, including irrigation and land drainage projects	Area of proposed work exceeds 1 hectare
3. Intensive livestock installations (not Annex I Projects)	Area of proposed new floor space exceeds 500 square metres
4. Reclamation of land from the sea	All development

As explained above, the fact that a project exceeds one of these thresholds means that it is highly likely that an environmental impact assessment will be required. Where the project does not exceed the thresholds then planning officers should consider the additional criteria mentioned above and require an environmental impact assessment if they believe that the project is likely to have a significant impact upon the environment.

---

195   See the Town and Country Planning (Environmental Impact Assessment) (England and Wales) Regulations 1999, SI 1999/293, the Environmental Impact Assessment (Scotland) Regulations 1999, SSI 1999/1 and the Planning (Environmental Impact Assessment) Regulations (Northern Ireland) 1999, SR 1999/73 as amended by SR 2008/17.

Traditionally, agricultural land use has enjoyed wide exclusions from development control in the United Kingdom, so that under section 55(2) (e) of the Town and Country Planning Act 1990, 'the use of any land for the purposes of agriculture or forestry ... and the use for any of those purposes of any building occupied together with land so used' is not considered to be development and, therefore, does not require planning permission. These exclusions posed a particular problem for the application of environmental impact assessment measures to many agricultural projects, since a number of Annex II developments would be excluded from the scope of development control. Secondary legislation has, therefore, been developed to deal with such loopholes. For example, the Environmental Impact Assessment (Agriculture) (England) Regulations (No.2) 2006 require a person wishing to restructure rural-land holdings, or to use uncultivated land or semi-natural areas for intensive agriculture, to first obtain a screening decision, from Natural England,[196] as to whether the proposed work is likely to have a significant effect upon the environment.[197] Should that decision confirm that this is the case then the work may only be conducted with the consent of the Secretary of State.[198] Equally, similar provision is made in respect of water-management projects in agriculture under the Water Resources (Environmental Impact Assessment) (England and Wales) Regulations 2003.[199]

Integrated Pollution Prevention and Control

Directive 96/61 on Integrated Pollution Control highlights a more holistic approach to environmental protection.[200] Where the European Community had, previously, sought to govern emissions to individual media (air, land and water) through separate controls, this Directive regulated all emissions that came from commercial sites. The principal aim of the Directive is to achieve a high degree of

---

196   In Scotland or Wales from the Scottish and Welsh Ministers respectively and in Northern Ireland from the Department of the Environment.

197   SI 2006/2522. Similar provision is made in Scotland, Wales and Northern Ireland under the Environmental Impact Assessment (Agriculture) (Scotland) Regulations 2006, SSI2006/582, the Environmental Impact Assessment (Agriculture) (Wales) Regulations 2007, WSI 2007/2933 (W.253) and the Environmental Impact Assessment (Agriculture) (Northern Ireland) Regulations 2007, SR 2007/421.

198   In Scotland the Scottish Ministers, in Wales the National Assembly and in Northern Ireland the Department for the Environment.

199   SI2003/164 as amended by the Water Resources (Environmental Impact Assessment) (England and Wales) (Amendment) Regulations 2006, SI 2006/3124. See also the Environmental Impact Assessment (Water Management) (Scotland) Regulations 2003, SSI2003/341, the Water Resources (Environmental Impact Assessment) Regulations (Northern Ireland) 2005 SR2005/32 as amended by the Water Resources (Environmental Impact Assessment) (Amendment) Regulations (Northern Ireland) 2006, SR2006/483.

200   [1996] L257/1.

environmental protection through preventing or, if this is not practicable, reducing emissions from these sites.[201] The Directive, however, has only limited relevance to the agricultural industry. It only applies to the particular types of commercial installations that are identified in an Annex to the Directive.[202] These installations include slaughter houses,[203] food production plants,[204] milk processing and treatment centres[205] and those disposing of or recycling animal by-products.[206] However, in terms of agriculture itself, only installations for the intensive rearing of poultry or pigs are identified within the Directive, and only if they have more than:[207]

- 40,000 places for poultry
- 2,000 places for production pigs (over 30 kg)
- or 750 places for sows

The Directive provided that no new installation could begin operation without first obtaining a permit.[208] Similarly, Member States were also required to oblige existing installations to have operating permits no later than 8 years after the Directive took legal effect – effectively by October 2004.[209] These permits, issued by the Environment Agency in England and Wales, the Scottish Environmental Protection Agency and by the Northern Ireland Environment Agency,[210] set emission limits for each of the pollutants likely to be emitted by the installation in significant quantities and also stipulated measures necessary to protect all environmental media and concerning the management of waste generated by the installation.[211]

---

201   *Ibid.* Article 1.

202   *Ibid.* Article 1 and Annex I.

203   *Ibid.* where the slaughter house has a carcase production capacity of more than 50 tonnes per day.

204   *Ibid.* food production plants dealing with animal raw materials come within the Directive if they have a finished product capacity of 75 tonnes or more. Those dealing with plant raw materials will come within the Directive if, based upon a quarterly average, they have a finished product capacity of more than 300 tonnes per day.

205   *Ibid.* those receiving more than 200 tonnes per day on an average annual basis come within the scope of the Directive.

206   *Ibid.* Installations with a treatment capacity in excess of 10 tonnes per day come within the scope of the Directive.

207   *Ibid.*

208   *Ibid.* Article 4.

209   *Ibid.* Article 5.

210   These permits are issued in the United Kingdom under the Environmental Permitting (England and Wales) Regulations 2007, SI 2007/3538, the Pollution Prevention and Control (Scotland) Regulations 2000, SSI 2000/323 and the Pollution, Prevention and Control Regulations (Northern Ireland) 2003.

211   *Ibid.* Article 9.

The operating permits must be based upon the emissions that could be achieved using 'Best Available Techniques' (BAT), whilst also taking into account the technical characteristics of each installation, its geographical location and the environmental conditions. BAT is defined by the Directive in the following terms:[212]

- *Best*: the most-effective technique through which to achieve a high level of protection for the environment as a whole
- *Available*: the techniques that have been developed on a sufficiently-large scale to allow them to be implemented by the relevant industrial sector, taking account of their economic cost and advantages, irrespective of whether they are produced within the Member State concerned, as long as they are reasonably accessible to the operator
- *Techniques*: the technology in question and the way in which the installation has been designed, built, maintained, operated or decommissioned

In applying the BAT criteria, account must also be taken of additional guidelines provided in the Directive,[213] and also 'best available technique' reference documents issued by the European Commission for each commercial sector covered by the Directive.[214] However, where an environmental quality standard requires it, even stricter emissions limits may be set than those required under the BAT principles.[215] This could, for example, arise where a stricter emissions limit was required to ensure that a body of water would meet the water quality standards set by the Water Framework Directive.

---

212   *Ibid.* Article 2(12).
213   *Ibid.* Annex IV.
214   *Ibid.* Article 16.
215   *Ibid.* Article 11.

Chapter 9
Agriculture and Food Safety

Consumer Confidence and Food Production

The food and drink industry plays a central role within the European economy. Its annual production is worth almost €600 billion, which amounts to approximately 15 per cent of the European Community's total manufacturing output.[1] It also provides employment for over 2.6 million people.[2] In recent years, however, the industry has experienced a series of food scares. Concerns about issues such as avian flu in chickens, eggs infected with salmonella and health risks posed by meat and, in the case of poultry, eggs obtained from animals fed on feedstuffs contaminated by dioxins, have all played a role in denting consumer confidence in food produce. Most recently, for example, concern that some pigs in Ireland had been given dioxin-contaminated feed led to the widespread withdrawal and destruction of Irish pork products.[3] Perhaps the greatest impact upon consumer confidence came with the Bovine Spongiform Encephalopathy (BSE) crisis that engulfed the European Community in the late 1980s and early 1990s.[4] The impact of this crisis was described by a European Parliament Committee of Inquiry in the following terms:[5]

> Since the beginning of the European Community, no debate has affected the daily life of individuals as much as this one. We must not underestimate the damage that the BSE crisis is causing amongst the general public, in particular the questioning of the food chain

For consumers, the BSE crisis caused major worries on two levels. Firstly, concern linked infected meat products with the human form of BSE, new variant

---

1   European Commission, *White Paper on Food Safety*, COM (1999) 719 final, 6.

2   *Ibid.*

3   *ENDS Report* January 2009, 5.

4   For analysis of the development of the BSE crisis see MacMaolain, C. (2007), *EU Food Law: Protecting Consumers and Health in a Common Market* (Oxford: Hart Publishing) 175–179 or Vincent, K. 'Mad Cows and Eurocrats- Community Responses to the BSE Crisis, [2004] 10 *European Law Journal* 500–504.

5   European Parliament Committee of Inquiry (Rapporteur Manuel Medina Ortega), *Report on alleged contraventions or maladministration in the implementation of Community law in relation to BSE, without prejudice to the jurisdiction of the Community and national courts,* (A4-0020/97) AI para. 1.4.

Creutzfeldt-Jakob Disease (NvCJD). Secondly, the fact that beef products were used in a wide variety of foods and medicines meant that a broad range of products came under suspicion.

The BSE crisis certainly caused major concern about the safety of the human food chain. It also caused the European Community to reflect, deeply, upon the adequacy of its own food safety laws and procedures. In the aftermath of the crisis a number of Community bodies published investigative reports. These highlighted failings by national governments, principally that of the United Kingdom, and by Community institutions such as the Council of the European Union and the European Commission.[6] They also identified deficiencies in the legal measures that had been available to deal with the crisis.[7] In turn, the European Commission carried out a major review of Community food safety laws. This led, first, to the publication of a green paper on the principles of food law within the European Community and then, subsequently, to the publication of a white paper on food safety.[8] These then culminated in the Community's adoption of Regulation 178/2002 establishing a general framework for the future development of EU food law.[9]

The Commission green paper noted that European Community safety laws had previously developed in an uncoordinated and piecemeal fashion. It also pointed to the need for core legislation setting out the fundamental principles upon which European Community Food Law should be based and the obligations that food producers should meet. The white paper called for consumers to be offered a wide range of safe and high quality products and for the European Community to re-establish public confidence in its food supply, food science and food controls.[10] This was to be achieved by adopting a comprehensive, integrated approach throughout the food chain, in which primary responsibility for food

---

6   See European Parliament Committee of Inquiry, (Rapporteur: Manuel Medina Ortega) *Report on alleged contraventions or maladministration in the implementation of Community law relating to BSE, without prejudice to the jurisdiction of the Community and national courts* (A4-0020/97) and European Parliament Committee of Inquiry, (Rapporteur: Reimer Böge) *Report on the European Commission's follow up of the recommendations made by the Committee of Inquiry into BSE* (A4-0362/97).

7   Court of Auditors, *Special Report 19/98 concerning the Community financing of certain measures taken as a result of the BSE crisis, accompanied by the replies of the Commission*, [1998] OJ C383/1 and Court of Auditors, Special Report 14/2001 *Follow up to Special Report No.19/98 on BSE, together with the Commission replies*, [2001] OJ C324/1.

8   European Commission, *Green Paper on the Principles of Food Law in the European Union*, COM (1997) 176 final and European Commission, *White Paper on Food Safety*, COM(1999) 719 final.

9   Regulation 178/2002 of the European Parliament and of the Council laying down the general principles and requirements of food law, establishing the European Food Safety Authority and laying down procedures in matters of food safety, [2002] OJ L31/1.

10   European Commission, footnote 8 above, 6–7.

safety rested with feed manufacturers, farmers and food operators. The white paper also, however, recognised the responsibility of consumers to ensure proper handling and storage.[11] These aspirations are clearly reflected in the Food Law Regulation, which emphasises that EC food law should adopt a 'farm to fork' approach. For example, the Regulation requires that all food and feed business operators, (at all stages of production, processing and distribution), should meet the requirements of food law.[12] Allied to this, the Regulation also introduces the principle of traceability. Feed businesses, farmers and food producers must all establish sufficient procedures to enable national food safety authorities to identify both their suppliers and also the businesses to which their own products have been supplied.[13] These suppliers include anyone who has supplied them with a food-producing animal or 'any substance intended to be, or expected to be, incorporated into a food or feed'.[14] Clearly, therefore, EC Food Law now directly encompasses the role that farmers play played in the process of food production.

The Food Regulation sets out three general objectives for Community food law:

1. Food law is to pursue one or more of the general objectives of achieving a high level of protection of human life and health and the protection of consumer interests, including fair practices in food trade, whilst taking account, where appropriate, of the protection of animal health and welfare, plant health and the environment[15]
2. Food law is to aim to achieve free movement of food and feed manufactured or marketed in accordance with the Regulation [16]
3. International standards that exist or are in contemplation must be taken into account in developing Community food laws 'except where such standards or relevant parts would be an ineffective or inappropriate means for the fulfilment of the legitimate objectives of food law or where there is scientific justification, or where they would result in a different level of protection from the one determined as appropriate in the Community'[17]

---

11   *Ibid.* 8.

12   Regulation 178/2002, Article 17, equally Article 3 provides that the food businesses covered by the Regulation include 'any undertaking, whether for profit or not and whether public or private, carrying out any of the activities related to any stage of the production, processing and distribution of food'.

13   *Ibid.* Article 18. Additionally food or feed placed upon the market, or which is likely to be placed upon the market, must also be adequately labelled or identified to facilitate traceability.

14   *Ibid.* Article 18(2).

15   *Ibid.* Article 5(1).

16   *Ibid.* Article 5(2).

17   *Ibid.* Article 5(3).

As one commentator has pointed out these three objectives neatly encapsulate the fundamental tension that exists between, on the one hand, the need to ensure food safety and, on the other hand, the importance of promoting trade in food products both within the Community and upon the broader world stage.[18]

The Food Regulation also establishes the fundamental food safety requirement that food that is unsafe should not be placed onto the market.[19] In practice, this encompasses food that is either injurious to health or unfit for human consumption.[20] Similarly, animal feed which is unsafe cannot be placed on the market or fed to any food-producing animal.[21] Feed will be considered to be unsafe if it will have an adverse effect on human or animal health or it makes food derived from farm animals unsafe for human consumption.[22]

Food Safety and the EC Treaty

Although the EC Treaty provided for the establishment of a Common Agricultural Policy and set out the objectives that this policy should pursue, it originally made no specific provision for food safety. The Treaty objectives for the Common Agricultural Policy provided, *inter alia*, for increased agricultural production levels, for farmers to achieve a fair standard of living and for consumers to pay reasonable prices.[23] However, no mention was made of food safety.

In practice, food is subject to European Community laws on free movement of goods, like every other product sold in the Community market place.[24] In the case of agricultural produce, the EC Treaty makes it clear that agricultural products are very much part of the common market and are subject to its rules, including those concerning the free movement of goods.[25] Similarly, the European Court of Justice has stressed that Community laws on free movement of goods form an integral part of each common organisation of the market established under the CAP.[26]

Despite the lack of any express reference to food safety issues within the EC Treaty, a range of food safety measures concerning agriculture have been adopted.

---

18   Cardwell, M. (2004), *The European Model of Agriculture* (Oxford: Oxford University Press) 266.

19   Regulation 178/2002, Article 14(1).

20   *Ibid.* Article 14(2).

21   *Ibid.* Article 15(1).

22   *Ibid.* Article 15(2).

23   See Article 33 of the EC Treaty.

24   For an examination of European Community law concerning the free movement of goods see Barnard, C. (2007), *The Substantive Law of the EU: The Four Freedoms*, 2nd ed., (Oxford: Oxford University Press) chapters 5–8.

25   See Articles 32(1) and 32(2) of the EC Treaty.

26   See variously Case 48/74 *Charmasson v Minister for Economic Affairs and Finance* [1974] ECR 1383; Case 83/78 *Pigs Marketing Board v Redmond* [1978] ECR 2347; Case 73/84 *Denkavit Futtermittel GmbH* [1985] ECR 1019.

These measures enable Member States to deny free movement to non-compliant products. For example, the Community adopted legislation on maximum pesticide residues in fruit and vegetables as early as 1976.[27] This legislation was predominantly based upon Article 37 of the Treaty, the treaty base for legislation concerning the Common Agricultural Policy. This use of Article 37 has also been supported by the European Court of Justice.The Court has found Article 37 to be an appropriate treaty base as it considers there to be a close connection between the protection of health and the general objectives of the CAP.[28]

> the protection of health contributes to the achievement of the objectives of the common agricultural policy laid down in Article [33(1)] of the Treaty, particularly where agricultural production is directly dependent on demand amongst consumers who are increasingly concerned to protect their health.

Today, the objectives of the Common Agricultural Policy still make no direct commitment to food safety. In practice, however, the broader EC Treaty does, today, make more specific provision for the European Community to act on food safety issues. For example, the 1986 Single European Act amended the EC Treaty to provide a more streamlined procedure for adopting legislation designed to approximate national laws in relation to the internal market.[29] In doing so it provided that Commission proposals for such legislation concerning health, safety, environmental protection and consumer protection should be based upon a high level of protection.[30] Subsequently, the 1993 Treaty on European Union also amended the EC Treaty to provide that the Community's activities should include making a contribution towards attaining a high level of health protection and also towards strengthening consumer protection.[31] This Treaty also introduced the requirement that all Community policies should ensure that their definition and implementation provides a high level of health protection.[32] Today, the European Community is also required to contribute towards protecting the health, safety and economic interests of consumers, in order to ensure a high level of consumer protection.[33]

---

27   By virtue of Council Directive 76/895, since repealed and replaced by Regulation 396/2005 of the European Parliament and of the Council on maximum residue levels of pesticides in products of plant and animal origin, [2005] OJ L70/, as amended by Commission Regulation 178/2006, [2006] OJ L29/3.

28   Case C-180/96 *United Kingdom v European Commission* [1996] ECR I-3903, para. 121. See also Case 68/86 *United Kingdom v Council* [1988] ECR 855, para. 12.

29   See Article 95 of the EC Treaty.

30   See Article 95(3) of the EC Treaty.

31   See Article 3(p) and 3(t) of the EC Treaty.

32   Article 152 of the EC Treaty.

33   Article 153 of the EC Treaty.

Looking beyond the EC Treaty, the Community institutions have now adopted a broad range of legislative measures designed to ensure that agricultural produce, and the foodstuffs they are incorporated into, pose no danger to public health. The Community legislation targets the agricultural produce itself. For example, maximum pesticide residue limits have been established, so that all foodstuffs intended for human or animal consumption must contain pesticide residue levels of less than 0.01 mg/kg.[34] Similarly, it has also acted to require farmers and food producers to ensure that food contaminants, including chemicals such as nitrate and heavy metals (for example, lead, cadmium, mercury), do not exceed levels that are toxicologically acceptable.[35] Additionally, Community legislation also controls the impact of materials that are used in agriculture. For example, the Community has banned the use of particular substances – those having a hormonal or thyrostatic effect and also beta-agonists, in animals intended for human consumption,[36] – on the basis that these substances 'may be dangerous for consumers and may also affect the quality of foodstuffs of animal origin'.[37] Additionally, Community law also controls the marketing and sale of pesticides and fertilisers.[38] In the case of fertilisers, these will only be approved for sale in the Community if they comply with a number of conditions. These include the requirement that 'under normal conditions of use they do not adversely affect human, animal or plant health and the environment'.[39]

---

34   Regulation 396/2005 of the European Parliament and of the Council on maximum residue levels of pesticides in products of plant and animal origin, [2005] OJ L70/1 as amended by Commission Regulation 178/2006, [2006] OJ L29/3. Article 18 establishes a default limit for pesticide residues at 0.01 mg/kg. The Regulation authorises the Community to establish more stringent residue limits for specific pesticides, but at the time of writing none had yet been established.

35   Council Regulation 315/93 laying down Community procedures for contaminants in foods, [1993] OJ L37/1, as amended by Commission Regulation 466/2001 setting maximum levels for certain contaminants in foodstuffs, [2001] OJ L77/1 as subsequently amended.

36   Council Directive 96/22 concerning the prohibition on the use in stock farming of certain substances having a hormonal or thyroststic action and of beta-agonists, [1996] OJ L125/1 as amended by Council Directive 2003/74, [2003] OJ L262/1.

37   *Ibid.* para. 3 to the preamble. This phrase also highlights the precautionary nature of the prohibition, which, as examined in Chapter 10, has been controversial in relation to international trade.

38   See Council Regulation 414/91 concerning the placing of plant protection products on the market, [1991] OJ L230/1 (as amended) and Regulation 2003/2003 of the European Parliament and of the Council relating to fertilizers, [2003] OJ L304/1.

39   *Ibid.* Regulation 2003/2003, Article 14.

The Institutional Structure of EC Food Safety Law

The 2002 Food Law Regulation also made provision for a new body, the European Food Safety Authority, ('the EFSA') which was to have an important role in assessing health risks associated with food and food products.

The establishment of the EFSA was one further consequence of the BSE crisis. In the aftermath of the crisis the systems of governance in place for the Community to deal with food safety issues came in for some criticism. The Court of Auditors, for example, observed that 'the occurrence of BSE and its aftermath have underlined the urgent need for the EU to design and to adopt a strategy... to deal, in a co-ordinated, effective and timely manner, with this kind of crisis'.[40] Failings occurred at both national and EU level. Chief amongst these were the fact that the European Commission had not had sufficient independent scientific advice and the fact that they had failed to do enough to ensure that Member States, particularly the United Kingdom, had taken adequate measures.[41]

The European Commission set out proposals for a European Food Safety Authority in its white paper.[42] It noted that this body would have a key role to play in both restoring and retaining consumer confidence in food safety.[43] It was to operate as an independent body, separate from the institutions and Member States. As to its role, the Commission defined this as being one of risk analysis.[44] In their view risk analysis had three elements, which, collectively, formed the three pillars of food safety:[45]

> *Risk assessment:* providing scientific advice based upon extensive information gathering and analysis
> *Risk management:* legislating on food safety issues and enforcing food safety laws
> *Risk communication:* keeping consumers informed and reducing the risk of unfounded food safety concerns arising

The Commission's white paper proposed that a European Food Safety Authority would have important roles in risk assessment and risk communication.[46] But it pointed to three reasons for not providing risk management functions to the EFSA. Firstly, the transfer of regulatory powers to an independent body would

---

40 Court of Auditors, *Special Report 19/98 concerning the Community financing of certain measures taken as a result of the BSE crisis, accompanied by the replies of the Commission*, [1998] OJ C383/1, para. 96.
41 See Vincent, K., footnote 4 above, 517.
42 European Commission, footnote 8 above, 14–21.
43 *Ibid.* 14.
44 *Ibid.* 14–15.
45 *Ibid.* 8.
46 *Ibid.*

amount to 'an unwarranted dilution of democratic accountability'.[47] Secondly, the Commission should retain its regulation and control functions in order to be able to discharge its functions under the EC treaties. Thirdly, there was no provision in the EC Treaty for regulatory powers to be conferred upon an independent body.[48]

The EFSA established under Regulation 178/2002 largely reflects that which was proposed in the European Commission's White Paper. Certainly, as the European Commission had envisaged, its principal roles are in risk assessment and risk communication. The Regulation requires the EFSA to:[49]

> ...provide scientific advice and scientific and technical support for the Community legislation and policies in all fields that have a direct or indirect impact on food and feed safety. It shall provide independent information on all matters within these fields and communicate on risks.

In practice, the Regulation confers a broad range of tasks upon the EFSA.[50] These have been summarised by O'Rourke into 6 main functions:[51]

1. Provide independent scientific advice (at the request of the Commission, Member States, national food bodies or the European Parliament) to evaluate food safety risks
2. Collection and analysis of scientific data covering issues such as nutrition, dietary patterns, exposure, risks, etc. in order to monitor food safety in the EU and to underpin EU policy in the area of food safety and nutrition
3. Safety evaluations of dossiers put forward by industry for Community level approval of substances or processes such as additives, foods for particular uses, Genetically Modified Organisms and novel foods
4. Identification of emerging food safety risks
5. Support to the Commission in cases of food safety crisis/food safety scares
6. Communication to the general public of its scientific advice and risk assessment

To this list might also be added the task of networking, promoting co-operation with and between national food safety bodies across the Member States.[52] The Commission White Paper pointed to this as an important mechanism to ensure that the most effective use was made of available resources, with scientific expertise

---

47   *Ibid.* 15.

48   *Ibid.*

49   Regulation 178/2002, Article 22(2).

50   *Ibid.* Article 23(a) to (l).

51   O'Rourke, R. (2005), *European Food Law* 3rd ed. (London: Sweet and Maxwell Ltd) 195.

52   See Regulation 178/2002, Article 36.

being shared across the European Union and at an international level.[53] This will involve the EFSA in working closely with national bodies such as the Food Standards Agency, in the United Kingdom, the Food Safety Authority in Ireland and other similar national food agencies.

Within the EFSA, its scientific panels and scientific committee have a key role in relation to these tasks. These bodies are responsible for producing the scientific opinions that form the basis of much of the EFSA's work.[54] There are eight scientific panels, each composed of independent scientific experts, covering the following subject areas:[55]

- The Scientific Panel on food additives, flavourings, processing aids and materials in contact with food
- The Scientific Panel on additives and products or substances used in animal feed
- The Scientific Panel on plant health, plant protection products and their residues
- The Scientific Panel on genetically modified organisms
- The Scientific Panel on dietetic products, nutrition and allergies
- The Scientific Panel on biological hazards
- The Scientific Panel on contaminants in the food chain
- The Scientific Panel on animal health and welfare

In taking decisions, the panels are each required to act by a majority of their members and to record all minority opinions.[56] The Scientific Committee is made up of the chairs of each Scientific Panel and six, independent, scientific experts who do not belong to any scientific panel.[57] Its job is to co-ordinate the working procedures of the panels and to provide opinions on issues that transcend any one panel or which do not fall within the competence of any panel.[58]

In carrying out its functions, the EFSA is required to operate with the objective of contributing towards a high level of protection of human life and health, whilst also taking account animal health and welfare, plant health and the environment, in the context of the operation of the internal market.[59] It has been argued that the reference to the internal market within these goals demonstrates that, in reality, the Community's principal aim, in establishing the EFSA, has been to develop

---

53 European Commission, footnote 8 above, 19–20.
54 *Ibid.* Article 28(1).
55 *Ibid.* Article 28(4).
56 *Ibid.* Article 28(7).
57 *Ibid.* Article 28(3).
58 *Ibid.* Article 28(2).
59 Regulation 178/2002, Article 22(3).

consumer confidence in the operation of the internal market, rather than simply providing greater levels of consumer protection.[60]

The EFSA should be distinguished from the European Community's Food and Veterinary Office, which was set up in April 1997.[61] The Food and Veterinary Office operates as an oversight body, with the role of ensuring that European Community food hygiene, veterinary and plant health legislation is being complied with at every stage of the food chain, in both Member States and other countries that export to the European Community. As such, therefore, the principal role of the Food and Veterinary Office is to carry out audits and inspections at food producers and to publish reports on its findings.[62]

The European Food Safety Authority's Role in the Development of Food Laws

The EFSA could easily be dismissed as being a relatively weak body, created simply to provide advisory opinions and to communicate with the public on food safety issues. However, as other commentators have pointed out, this would not be an accurate reflection of its role.[63] Although the EFSA lacks any formal regulatory powers its scientific opinions will, in practice, have a strong influence upon the development of Food Law in the European Community.

The Food Regulation provides that, in order to achieve a high level of protection for human health and life, Community food laws should, normally, be based upon risk analysis.[64] Additionally, it also provides that the EFSA's scientific opinions are to serve as the scientific basis for drafting and adopting Community food laws.[65] In effect, such provisions mean that, whilst there may be no legal obligation upon the Community institutions to follow an EFSA opinion, it will be likely to have a major influence upon Community legislators.[66] However, it will not necessarily always be determinative as Article 6(3) of the Food Regulation also authorises Community legislators to take into account 'other factors legitimate to the matter under consideration'. The preamble to the Regulation provides that such factors can include 'societal, economic, traditional, ethical, and environmental factors' along with the feasibility of controls.[67]

---

60   Kanska, K., Wolves in the Clothing of Sheep? The Case of the European Food Safety Authority, [2004] *European Law Review* 711–727.

61   For further information on the Food and Veterinary Office see http://ec.europa. eu/food/fvo/what_en.htm.

62   See O'Rourke, R., footnote 51 above, 97.

63   See, Chalmers, D., 'Food for Thought': Reconciling European Risks and Traditional Ways of Life, [2003] 66 *Modern Law Review*, 532–562 and Kanska, K., Wolves in the Clothing of Sheep? The Case of the European Food Safety Authority, [2004] 29 *European Law Review* 711–727.

64   Regulation 178/2002, Article 6(1).

65   *Ibid.* Article 22(6). See also Articles 6(2) and 6(3).

66   See Kanska, K., footnote 63 above.

67   Regulation 178/2002, recital 19 to the preamble.

The EFSA's scientific opinions will also have an impact upon national laws within the Member States. As a general principle, where there is no specific Community law, food is deemed to be safe if it conforms to the national food laws of the Member State in whose territory it is marketed.[68] Equally, as noted above, food law is to be based on risk analysis, in order to secure the high level of protection of human health and life required by the Food Regulation.[69] The Regulation makes it clear that references to 'food law' encompass both Community and national laws.[70] Equally, it specifically provides for legislators to take the EFSA's opinions into account. Therefore, at a national level, these legislators will also have to justify any derogation from the risk assessment contained within EFSA scientific opinions.[71] The net result of this situation is that the operation of the internal market has been strengthened, since it is more difficult for Member States to justify restrictions upon the free movement of food.[72]

Finally, the EFSA's scientific opinions will also have an important impact upon private parties in national courts.[73] Although not obliged to do so, national courts are likely to give weight to these opinions when assessing culpability for the marketing or sale of unsafe food. As Chalmers notes, 'if the EFSA has issued an opinion suggesting that a product is not safe, it would be impossible for a producer to use this defence'.[74]

Responding to Food Safety Emergencies: The RAPID Alert System

The BSE crisis exposed inadequacies within the European Community's procedures for dealing with food safety emergencies.[75] The Community had originally introduced a rapid alert system in 1992, which required Member States to inform the European Commission of any emergency measures that they adopted in order to restrict the marketing of unsafe food products.[76] However, the Commission white paper on food safety proposed that the EFSA should have a central role within a re-launched rapid alert system.[77] This re-launched system was put in place by the Food Regulation.

---

68 Regulation 178/2002, Article 14 (9).
69 *Ibid.* Article 6(1).
70 *Ibid.* Article 3(1).
71 Chalmers, D., footnote 63 above, 542.
72 Kanska, K., footnote 63 above, 711.
73 Chalmers, D., footnote 63 above, 542.
74 *Ibid.*
75 See O'Rourke, R., footnote 51 above, 28. See also the preamble to the Food Regulation 2002, paragraphs 60–61.
76 See Council Directive 92/59/EEC, ([1992] OJ L228/24) Article 8.
77 European Commission, footnote 1 above, 19.

The Food Regulation puts in place a food safety network, comprised of an identified contact point designated by each Member State, the European Commission and the EFSA.[78] Additionally, the Regulation also enables applicant countries, third countries or international organisations to be included within the rapid alert system, through agreements with the Community.[79] When any member of the network has any information about a serious direct or indirect risk to human health, deriving from food or animal feed, they are required to immediately notify that information to the European Commission, which will then notify the other members of the network.[80] EFSA can also supplement this with any scientific or technical information that it may have that could help to facilitate rapid, appropriate, risk management action by the Member States.[81] As a general rule, information on product identification, the nature of the risk to human health posed by the product and the protective measures that have been taken should also be made available to the general public.[82]

Under the Food Regulation, Member States are required to immediately notify the European Commission, under the rapid alert system, of:[83]

- Any measure they adopt which is aimed at restricting the placing on the market or forcing the withdrawal from the market or the recall of food or feed in order to protect human health and requiring rapid action
- Any recommendation or agreement with professional operators which is aimed, on a voluntary or obligatory basis, at preventing, limiting or imposing specific conditions on the placing on the market or the eventual use of food or feed on account of a serious risk to human health requiring rapid action
- Any rejection, related to a direct or indirect risk to human health, of a batch, container or cargo of food or feed by a competent authority at a border post within the European Union

In each case, the notification must be accompanied by detailed information setting out the reasons for taking this action.[84] All of this information is then forwarded to other members of the rapid alert network.[85]

---

78   Regulation 178/2002, Article 50(1).
79   *Ibid.* Article 50(6).
80   *Ibid.* Article 50(2).
81   *Ibid.* See also Article 35.
82   *Ibid.* Article 52.
83   *Ibid.* Article 50(3).
84   *Ibid.* It may also be followed by supplementary information, in particular should the measures taken by that Member State be modified or withdrawn.
85   *Ibid.* Any supplementary information will also, subsequently, be notified to other network members.

Additionally, when food or animal feed is rejected at a Community border post, the European Commission is also required to notify all border posts as well as the third country from which the goods originated.[86]

For their part, Member States must immediately inform the European Commission of any action that they take, or measures that they implement, as a result of the information circulated within the rapid alert system. Details of this action will also be passed to other network members.

Where action by one or more Member States, under the above measures, is unlikely to be sufficient to contain a serious risk to human health, animal health or the environment, then the European Commission has the power to take co-ordinated action at a Community level, either on its own initiative or at the request of a Member State.[87]

Where the risk is posed by food or animal feed that originated within a Member State the Commission may:[88]

- Suspend the placing on the market or use of the food concerned
- Suspend the placing on the market or use of the animal feed concerned
- Lay down special conditions for the food or feed concerned
- Adopt any other appropriate interim measure

Where the risk is derived from food or animal feed imported from third countries then the Commission has the power to:[89]

- Suspend imports of the food or animal feed concerned from all or part of the third country concerned and, if applicable, from a third country of transit
- Lay down special conditions for the food or animal feed concerned from all or part of the third country concerned
- Adopt any other appropriate interim measure

In order to take such action, the European Commission is required to act in accordance with the regulatory committee procedure established in the Council's comitology decision.[90] This requires the Commission to act under the supervision of a regulatory committee established by the Council.[91] In emergencies, however, it can take provisional measures, but, within 10 working days, those provisional

---

86   *Ibid.*

87   *Ibid.* Article 53(1).

88   *Ibid.*

89   *Ibid.*

90   *Ibid.* Article 53(1) and Article 58(2). The 'comitology decision' refers to Council Decision 1999/468 laying down the procedures for the exercise of implementing powers conferred upon the Commission, [1999] OJ L184/23.

91   Under Article 5 of Council Decision 1999/468.

measures must be confirmed, amended, revoked or extended under the regulatory committee procedure.

Finally, the Food Regulation requires the European Commission, acting in close co-operation with the EFSA and Member States, to put in place a crisis management plan, which identifies the types of situation that could arise in which direct or indirect risks to human health would not be adequately controlled by the measures examined above.[92] The crisis management plan must specify the practical procedures that would be necessary to manage such a crisis, including the principles of transparency to be applied and a communication strategy.[93] Where the type of crisis situation envisaged by the crisis management plan does actually occur, the Commission is then required to immediately establish a crisis management unit, with the EFSA participation to provide scientific and technical assistance.[94] The role of the crisis unit is to collect and evaluate all relevant information and identify the options available to prevent, eliminate, or reduce to an acceptable level, the risks to human health.[95] In doing so it can ask for assistance from any public or private person whose expertise, it believes, will be necessary to enable it to manage the crisis effectively.[96] It also has an obligation to keep the public informed of the risks involved and of the measures that have been taken to counter them.[97]

Risk and Regulation: A Precautionary Approach

As has been observed, the EC's Food Regulation creates a two step process for risk regulation. The EFSA has the role of scientifically evaluating risks to human health posed by food products and of communicating its views to the EC institutions, whilst it is those EC institutions that will then ultimately adopt any legislation. But what should happen when the EFSA's scientific reports reveal scientific uncertainty, merely showing that a product 'may' pose a risk to human health? In this situation the Food Regulation authorises Community legislators to act on the basis of the 'precautionary principle'.[98]

The 'precautionary principle' enables action to be taken to eliminate potential dangers even before a causal link has been established by clear scientific evidence.[99] The principle is derived from the *Vorsorgeprinzip*, which influenced German

---

92 Regulation 178/2002, Article 55.

93 *Ibid.*

94 *Ibid.* Article 56.

95 *Ibid.* Article 57.

96 *Ibid.* Article 57(2).

97 *Ibid.* Article 57(3).

98 Regulation 178/2002, Articles 6(3) and 7.

99 Freestone, D., and Ryland, D., EC Environmental Law After Maastricht, [1994] 45 *Northern Ireland Legal Quarterly* 156.

environmental law in the 1970s and 1980s.[100] The principle was, subsequently, introduced into the EC Treaty by the Treaty on European Union, as one of the core principles that EU environmental policy should be based upon.[101]

Although the EC Treaty, today, continues to refer to the 'precautionary principle' merely in relation to EC environmental policy, it has also now become a core aspect of European Community food law.[102] It was the European Court of Justice which first extended the scope of the principle to include human health,[103] noting that Community measures to prevent the spread of BSE could, legitimately, be adopted before the seriousness of any risk to human health had been fully proven.[104] Its inclusion in the Food Regulation, as an element of risk management, now cements this position. Indeed the Court of First Instance has since identified the 'precautionary principle' as being a general principle of EC law, noting that:[105]

> The precautionary principle can be defined as a general principle of Community law requiring the competent authorities to take appropriate measures to prevent specific potential risks to public health, safety and the environment, by giving precedence to the requirements related to the protection of those interests over economic interests.

The EC treaty does not provide any definition of the 'precautionary principle', whilst the Food Regulation provides merely that 'where ... the possibility of harmful effects on health is identified but scientific uncertainty persists',[106] provisional measures may be taken. This begs the question, how harmful must such effects be likely to be? There are two potential answers to this question, depending upon the manner in which the 'precautionary principle' is interpreted.

A strong interpretation of the 'precautionary principle' would, basically, require a reversal of the scientific burden of proof in all cases in which there was potential for environmental damage or damage to human health. Scientific proof would be

---

100   Boehmer-Christiansen, S., The Precautionary Principle in Germany – enabling Government, in O'Riordan, T., and Cameron J., (eds) (2004), *Interpreting the Precautionary Principle* (London: Earthscan Publications).

101   Today see Article 174(2) of the EC Treaty.

102   See Berends G., Carreno, I., Safeguards in Food Law – Ensuring Food Scares are Scarce, [2005] 30 *European Law Review* 399, Streinz, R., The Precautionary Principle in Food Law, [1998] 4 *European Food Law Review* 413.

103   Case C-18/1996 *United Kingdom v European Commission* [1998] ECR I-2265 para. 99 and Case C-157/1996 *R v Ministry of Agriculture, Fisheries and Food, ex. parte National Farmers Union* [1998] ECR I-2211 para. 63.

104   See MacMaolain, C. (2007), *EU Food Law: Protecting Consumers and Health in a Common Market*, (Oxford: Hart Publishing) 196.

105   See Joined Cases T-74/00, T-76/00 and T141/00 *Artegodan & Others v European Commission* [2002] ECR II-4945, para. 184.

106   Regulation 178/2002, Article 7(1).

required to establish that the product concerned would not cause this damage.[107] This would, essentially, amount to a 'zero risk' policy, an approach that has been rejected by both the European Commission and the Court of First Instance.[108] In the context of a review of the legality of Community law banning particular antibiotics in animal feeds, the Court of First Instance noted that:[109]

> a zero risk does not exist, since it is not possible to prove scientifically that there is no current or future risk associated with the addition of antibiotics to feedstuffs.

Alternatively, a weak interpretation would require that the principle should only apply in situations where there was a possibility of significant damage to the environment or to human health. This, obviously, requires decision makers to consider the degree of risk that will be accepted before the principle is applied. Support for this weaker interpretation can also be gleaned from European Community sources. In 2000, the European Commission issued a *Communication on the Precautionary Principle*,[110] which the Court of First Instance has referred to as setting out 'the current state of the law' on the principle.[111] In its communication, the Commission referred to the principle being used in relation to the 'potentially dangerous effects' of a phenomenon, product or process and on the basis of an analysis of the costs and benefits of action or lack of action.[112] This suggests that the Community should consider both the magnitude of harm and also the cost effectiveness of measures before taking action.[113]

This leaves one issue, how much scientific knowledge must there be of the risk to the environment or human health? In *Pfizer*, the Court rejected arguments that the Council of the European Union should only have been able to rely upon the 'precautionary principle' to adopt legislation banning particular antibiotics in animal feed if they could demonstrate a probable risk. It noted that the Community institutions did not have to wait until 'the first infection, first colonisation or the

---

107   See Lee, M. (2005), EU *Environmental Law: Challenges, Change and Decision Making* (Oxford: Hart Publishing) 100; Haigh, N., The Precautionary Principle: The Commission's View, [1996] 14 *Environmental Liability* 55 or MacDonald, J., Appreciating the Precautionary Principle as an Ethical Evolution in Ocean Management, [1995] 26 *Ocean Development and International Law* 263.

108   See, respectively, European Commission, *Communication on the Precautionary Principle*, COM(2000) 1 Final, 9 and Case T-13/99 *Pfizer Animal Health SA v Council of the European Union* [2002] ECR II 3305, para. 145.

109   *Ibid.*

110   European Commission, footnote 108 above.

111   Case T-13/99 *Pfizer Animal Health SA v Council of the European Union*, footnote 108 above, para. 123.

112   European Commission, footnote 108 above.

113   See also Lee, M., footnote 107 above, 98.

first proof of transfer in a human' provided clear evidence'.[114] Equally, however, the Court also made it clear that the principle did not enable legislation to be adopted purely on the basis of hypothetical risks identified by 'mere conjecture which has not been scientifically verified'.[115] Instead, the middle ground, supported by the Court, was that precautionary measures can be adopted when scientific data supports the existence of the risk even though 'the reality and extent thereof have not been 'fully' demonstrated by conclusive scientific evidence'.[116] The Court concluded that:[117]

> The scientific risk assessment must enable the competent public authority to ascertain, on the basis of the best available scientific data and the most recent results of international research, whether matters have gone beyond the level of risk that it deems acceptable for society... That is the basis on which the authority must decide whether preventive measures are called for.

Any food safety measures adopted on the basis of the 'precautionary principle' are, however, only provisional.[118] The Food Regulation requires that they be reviewed within a reasonable period of time, which is to be determined by 'the nature of the risk to life or health identified and the type of scientific information needed to clarify the scientific uncertainty and to conduct a more comprehensive risk assessment'.[119] Additionally, any measures based upon the 'precautionary principle' must also be proportionate and no more restrictive to trade than required to achieve the high level of health protection required by the Community.[120]

The preamble to the Food Regulation provides that one additional objective of the Regulation is to ensure that the 'precautionary principle' is applied on a uniform basis throughout the Community, to avoid creating barriers to trade.[121] This is an issue that had previously come before the Court in *European Commission v France*.[122] The case concerned France's refusal to respect a Commission decision authorising beef exports from the United Kingdom following the end of the BSE outbreak. France justified its refusal to allow the import of this beef on the basis

---

114 Case T-13/99 *Pfizer Animal Health SA v Council of the European Union*, footnote 108 above, para. 179.

115 *Ibid.* para. 143. See also Case T-70/99 *Alpharma Inc. v Council of the European Union* [2002] ECR II-3305, para. 156.

116 *Ibid.* para. 144. See also Case T-70/99 *Alpharma Inc.* footnote 168 above, para. 157.

117 *Ibid.* para. 162. This also reflects the position that the European Commission had adopted in its communication on the 'precautionary principle', see European Commission, footnote 108 above, 13.

118 Regulation 178/2002, Article 7(1).

119 *Ibid.* Article 7(2).

120 *Ibid.*

121 *Ibid.* recital 20.

122 Case C-1/00 *European Commission v France* [2001] ECR I-9989.

of the opinions of its own scientific advisory committee, which disagreed with
the conclusions reached by the European Community's own scientific committee.
France argued, on the basis of their scientific evidence, that a precautionary
approach to BSE required the continuation of a ban on exports of beef from the
United Kingdom. The European Commission argued strongly that no one Member
State could rely upon the scientific opinion of a national body to substitute its own
assessment of the risks for that relied upon by the Commission.[123] Although the
Commission's argument was endorsed by Advocate General Mischo, the Court
did not directly deal with the point.[124] Instead, the Court simply noted that Member
States could not allege the unlawfulness of a decision, addressed to it as a defence,
in infringement proceedings for not implementing that decision.[125] In effect, if it
wanted to do so, France should have challenged that validity by bringing judicial
review proceedings against the Commission under Article 230 EC.

Genetically Modified Organisms

In the absence of Community measures to harmonise national laws, concerning
the use of genetically modified organisms (GMOs), it was likely that differing
national restrictions on trade, would develop, justified on the basis of the potential
harm posed by GMOs to the environment and human health. At the same time, the
question of the degree of harm that GMOs actually do pose to the environment
and human health is often shrouded in scientific uncertainty. Consequently, the
'precautionary principle' also has had an important role to play in the development
of EC legislation concerning GMOs.

GMOs are legally defined as being 'organisms, with the exception of human
beings, in which the genetic material has been altered in a way that does not occur
naturally by mating and or natural recombination'.[126] Their use in agriculture has
been highly controversial within the European Community. Surveys of public
opinion across the Community have shown foods containing GMOs to be highly
unpopular with consumers.[127] This unpopularity often stems from a mistrust of
scientific research. The European Community has sought to address consumer

---

123   *Ibid.* paragraph 88 of the Court's judgment.

124   *Ibid.* see paragraph 120 of the Advocate General's opinion.

125   *Ibid.* see paragraph 101 of the Court's judgment.

126   European Parliament and Council Directive 2001/18/EC on the deliberate release
into the environment of GMOs and repealing Council Directive 90/220/EEC, [2001] OJ
L106/1, Article 2.

127   See generally Tsioumani, E., Genetically Modified Organisms in the EU: Public
Attitudes and Regulatory Developments, [2004] 13 *Review of European Community and
International Environmental Law* 279, Cardwell, M., The Release of Genetically Modified
Organisms into the Environment: Public Concerns and Regulatory Responses, [2002]
Environmental Law Review 156 or Burton, M., et al., Consumer Attitudes to Genetically
Modified Organisms in the UK, [2001] 28 *European Review of Agricultural Economics*

concerns by putting in place traceability measures and labelling requirements. 'Traceability' is defined as 'the ability to trace GMOs and products produced from GMOs at all stages of their placing on the market through the production and distribution chains.'[128] The principle requires the operator who first places products containing or consisting of GMOs onto the market to give written notice to any person who acquires them from him of the fact that they contain or consist of GMOs.[129] This obligation is then repeated at each subsequent stage of the food production and distribution process. Ultimately, when the product is offered for sale, consumers are also protected by labelling requirements. Pre-packaged products consisting of or containing more than a threshold level of GMO (where more than 0.9 per cent of the food or of an individual food ingredient consists of GMO or has been produced from GMO, and that GMO presence is 'adventitious' or 'technically unavoidable'),[130] must carry a label stating either that 'this product contains genetically modified organisms' or using an amended version of this phrase specifying the particular genetically modified organism.[131] In the case of non-packaged goods, this warning must be incorporated into the product's display.[132] The label must also display a unique numerical identifier that is allocated to the GMO following its authorisation.[133] Questions, however, have been raised as to the adequacy of these labelling requirements. Although they apply to food containing or derived 'from' a GMO, they do not apply to food manufactured 'with' a GMO, such as GMO-based processing aid.[134] Equally, they do not apply to agricultural produce obtained from livestock reared on GMO based feed.[135]

---

479. Carson L., Lee R., Consumer Sovereignty and the Regulatory History of the European Market for Genetically Modified Foods [2005] 7 *Environmental Law Review* 173.

128 Regulation 1830/2003 of the European Parliament and the Council concerning the traceability and labelling of GMOs and the traceability of food and feed products produced from GMOs. [2003] OJ L268/24. Article 3.

129 *Ibid.* Article 4(1). The written notification must also contain the unique identifier assigned to the GMOs in question under Community Law (In relation to this identifier see Article 8 of the Regulation and also Commission Regulation 65/2004 establishing a system for the development and assignment of unique identifiers for genetically modified organisms, [2004] OJ L10/5.

130 See Deliberate Release Directive Articles 12a and 21 GM Food and Feed Regulation, Article 12(2).

131 Regulation 1830/2003, Article 6.

132 *Ibid.*

133 In relation to this unique identifier see Commission Regulation 65/2004 establishing a system for the development and assignment of unique identifiers for genetically modified organisms, [2004] OJ L10/5.

134 Carson, L., Lee R., footnote 127 above, 183.

135 *Ibid.* 184. See also MacMaolain, C., The New Genetically Modified Food Labelling Requirements: Finally a Lasting Solution? [2003] 28 *European Law Review* 865.

Consumers are not just concerned about the impact that GMOs may have on human health or, indeed, upon the broader environment, they are also driven by broader concerns, such as the probity of interfering with nature or the social concern about the central role that GMOs provide to commercial interests within agricultural production.[136] This has created a highly polarised debate within the European Community, which was clearly visible in a de facto moratorium that existed on the authorisation of GMOs between 1998 and 2004. During this period, Member States, with sufficient votes to create a blocking minority in the Council of the European Union, indicated that they would reject proposals to authorise new GMOs until they felt that the European Community had established a satisfactory legal framework on the labelling and traceability of GM products.[137] In the light of Member State concerns and broader public opinion the European Community developed its regulatory framework. This now controls both the placing of GM products onto the market and also the presence of GMOs in food products. Today, this legal framework is principally based upon the following measures:[138]

- Directive 2001/18/EC on the deliberate release, into the environment, of genetically modified organisms, (the 'Deliberate Release Directive')[139]
- Regulation 1829/2003 on genetically modified food and feed, (The 'GM Food and Feed Regulation')[140]
- Regulation 1830/2003 concerning the traceability and labelling of genetically modified organisms and of food and feed products produced from genetically modified organisms (the 'Labelling and Traceability Regulation')[141]

Unfortunately, in adopting these measures, the European Community showed little regard for its own better regulation agenda. Instead, they have created an unduly complex matrix of overlapping provisions. However, collectively, the legislation establishes a twofold approach. On the one hand, it prevents the unauthorised use of GMOs. On the other hand, it introduces the labelling and traceability requirements mentioned above.

The Deliberate Release Directive and the GM Food and Feed Regulation are both relevant to agricultural production. Taken at face value it might seem that it

---

136   See M. Lee, footnote 107 above, 250.

137   See generally E. Tsioumani, footnote 127 above, 282. The Member States concerned were Denmark, Greece, France, Italy and Luxembourg. Subsequently, following EC expansion they were also joined by Austria.

138   See also Commission Regulation 65/2004 establishing a system for the development and assignment of unique identifiers for genetically modified organisms, [2004] OJ L10/5.

139   [2001] OJ L106/1.

140   [2003] OJ L268/1.

141   [2003] OJ L268/24.

is the Deliberate Release Directive that is most directly applicable to agriculture, applying, as it does, to the placing on the market of GMOs or products containing GMOs or to the deliberate release of GMOs into the environment.[142] In contrast, the GM Food and Feed Regulation applies to GMOs intended for use as food and animal feed , to food and feed containing or consisting of GMOs and also to food and feed produced from or containing ingredients produced from GMOs'.[143] Each measure sets out different authorisation procedures, each requiring applications to provide comprehensive documentation. The authorisation procedures established under the Directive predate the European Community's Food Regulation and the establishment of the EFSA. Consequently, they rely principally upon national food safety bodies. In contrast, the Regulation was introduced after the introduction of the Food Regulation and provides a prominent role for the EFSA. In practice, as most GMOs used in agriculture have been developed for use in food, considerable overlap exists between these procedures. As a result, single application can be made, under the GM Food and Feed Regulation, to obtain authorisation for the use of the GMO in agriculture and also for its use in food and/or animal feed.[144]

Under the GM Food and Feed Regulation, GMOs used in food must not have adverse effects on human health, animal health or the environment, must not mislead the consumer and must not differ from the food which it is intended to replace, to such an extent that its normal consumption would be nutritionally disadvantageous to the consumer.[145] The Regulation establishes a quite centralised procedure, giving the EFSA the principal task of assessing applications against these criteria. This addresses one of the central problems with risk assessments conducted by national bodies – the danger that other Member States will fail to accept their probity.[146] The initial application is, again, made to a competent national authority within a Member State, (DEFRA in the United Kingdom), containing the technical information required by the Regulation.[147] Where the application relates to GMOs or food containing or consisting of GMOs, the applicant must also supply the technical information required in the Deliberate Release Directive, including

142   Article 1 provides that the Directive applies to the deliberate release of GMOs into the environment and to placing GMOs on to the market as or in products.

143   Regulation 1829/2003, Articles 3 and 15.

144   *Ibid.* Article 5(5).

145   *Ibid.* Article 4(1). Similar criteria are laid down by Article 16 in relation to animal feed. For animal feed an additional criterion requires that the feed must not harm or mislead consumers by impairing the distinctive features of animal products. Additionally the final criterion requires that consumption should not be nutritionally disadvantageous for animals or humans.

146   See Hervey, T., Regulation of Genetically Modified Products in a Multi-Level System of Governance: Science or Citizens, (2001) 10 *Review of European Community and International Environmental Law* 321 or Lee, M., footnote 107 above, 245.

147   Regulation 1829/2003, Article 5(2). As in the case of the Deliberate Release Directive, the information which must be accompanied with the application is prescribed.

an environmental risk assessment.[148] The national authority must acknowledge the application, in writing, within 14 days, inform the EFSA, without delay, and make the application and any additional information supplied by the applicant available to the EFSA.[149] The EFSA, in turn, must then inform the other Member States and the Commission, make the application and supporting documentation available to them and make a summary of the application available to the public.[150] The EFSA is required to 'endeavour' to provide an opinion on the application within 6 months of the application.[151] Before providing this opinion the EFSA may, also, where the application relates to the deliberate release or placing on the market of GMOs, ask a competent national authority to carry out an environmental risk assessment.[152] It must ask for such an environmental risk assessment to be conducted when the application relates to GMOs that are to be used as seeds or other plant-propagating materials.[153] Ultimately, the EFSA must forward its opinion to the European Commission, the Member States and the applicant, setting out, also, its reasons for reaching that opinion, the information upon which it is based and the opinions of any competent national authorities that have been consulted.[154] The European Commission is then required, within three months of receiving the EFSA's opinion, to submit a draft decision on the application to the Standing Committee on the Food Chain and Animal Health.[155] This is a regulatory committee established under the comitology decision. If the committee, acting by qualified majority, approves the decision then the Commission can adopt it. Should this not occur, the draft decision will be passed to the Council of the European Union, which can adopt or reject it by qualified majority. If the Council fails to achieve a qualified majority to adopt or reject the decision, then the Commission can proceed to adopt the decision. This latter situation occurred in 2004, when application had been made to approve a product for human consumption (it was not intended for use in agriculture).[156] The application had been made under an equivalent procedure contained in the Novel Foods Directive, which prior to the adoption of the GM Food and Feed Regulation, provided the regulatory base for the authorisation of

---

148   *Ibid.* Article 5(5).

149   *Ibid.*

150   *Ibid.*

151   *Ibid.* Article 6(1), though the Article does enable this time limit to be exceeded whenever the EFSA seeks supplementary information from an applicant.

152   *Ibid.* Article 6(3)(c).

153   *Ibid.*

154   *Ibid.* Article 6(6).

155   *Ibid.* Article 7. Under this Article, where the draft decision is not in accord with the opinion of the EFSA the European Commission must explain the reason. The comitology decision refers to Council Decision 99/468 laying down the procedures for the exercise of implementing powers conferred on the Commission, [1999] OJ L184/23.

156   See Tsioumani, E., footnote 127 above, 283.

food and food ingredients containing or produced from GMOs.[157] In this case, neither the Standing Committee nor the Council was able to reach agreement upon the application. The Commission then proceeded to adopt a decision approving the application. The fact that this decision was adopted even though only six Member States had supported the application in the Council raises concerns about the democratic accountability of the procedure, even if it did end the six-year moratorium on new GMO authorisations within the European Community.

In contrast, the earlier Deliberate Release Directive, which predated the establishment of the EFSA, placed Member State authorities at the forefront in deciding whether GMOs should be placed on the market or otherwise released into the environment. As noted above, however, most applications today will be made under the GM Food and Feed Regulation, since this provides a one-stop procedure covering both the deliberate release of GMOs and also their use in food and animal feed. The Deliberate Release Directive requires those seeking permission to apply to the relevant national authority in the Member State in which this is to occur (again DEFRA in the United Kingdom).[158] An environmental risk assessment, identifying and evaluating any potential adverse effects that the GMO may, directly, indirectly, immediately or delayed, have on human health and the environment, forms a central part of this application[159] The national body has 90 days to prepare an assessment report indicating whether it believes that the application should be approved, and if so under which conditions, or refused.[160] If the assessment report did not support granting the application then that application is deemed to have been rejected.[161] Copies of this report, together with the information relied upon in preparing it, are sent to the applicant and to the European Commission, which must forward it along with supporting information to the competent authorities of each other Member State.[162] If the assessment report supports the application then that application can be approved in one of two ways. Firstly, if no reasoned objection is made by another Member State authority or by the European Commission, within 60 days of the circulation of the assessment report.[163] Secondly, where there has been some objection, the application can still be approved if the objecting national food safety authorities and the Commission reach agreement on the

---

157   Council and European Parliament Regulation 258/97 on novel foods and food ingredients, [1997] OJ L43/1. Article 38 of Regulation 1829/2003 on genetically modified food and feed excludes the regulation of GMO food and food ingredients from the scope of the Novel Foods Directive.

158   Article 13. The application must be accompanied by a detailed technical dossier containing the information required by these Articles, which includes an environmental risk assessment.

159   Directive 2001/18, Article 13(2) (b) and Annex II.

160   *Ibid.* Article 14(2).

161   *Ibid.* Article 15(2).

162   *Ibid.* The European Commission must then circulate the report within 30 days of receipt.

163   *Ibid.* Article 15(3).

application within 105 days of the circulation of the assessment report.[164] In either circumstance, the national authority that received the application has to provide written consent to the applicant, and also notify the other Member States and the European Commission, within 30 days.[165] However, where an objection was raised and maintained by a Member State authority, the application would be referred to the Standing Committee on the Food Chain and Animal Heath.[166]

GMOs and the Precautionary Principle

Under both Community measures, the decision-making processes clearly adopt a case by case approach. The 'precautionary principle' will have an important role to play when the risks associated with each application are being analysed. In practice, the GM Food and Feed Regulation makes no express reference to the 'precautionary principle'. In contrast, its importance within the Deliberate Release Directive is explicit. Article 1 of the Directive stipulates that the aim of the Directive is to harmonise national laws in accordance with the 'precautionary principle'. Equally, Article 4(1) requires Member States to act in accordance with the principle in order to avoid any 'adverse effects on human health and the environment which may arise from the deliberate release or placing on the market of GMOs'. Finally, Annex II of the Directive requires Member States to ensure that the environmental risk assessment required under the Directive is conducted in accordance with the 'precautionary principle'. There are two likely reasons for these differing approaches. Firstly, that between the adoption of the Deliberate Release Directive and the GM Food and Feed Regulation the Court of First Instance had established that the 'precautionary principle' was a general principle of EC law.[167] Secondly, that during this period the Food Regulation had been adopted, providing for Community institutions to act on the basis of the principle when developing food law.[168] Consequently, the principle will, in practice, also have an important influence upon risk assessments made under the GM Food and Feed Regulation.

Once an assessment has been made of the risks posed by a particular GMO or product based upon that GMO, political decision makers will have to decide whether to authorise its use. Here the GM Food and Feed Regulation seems to adopt a very strict risk management policy It requires that GMOs 'must not have

---

164   *Ibid.* Articles 15(1) and Article 15(3).

165   *Ibid.* Article 15(3).

166   *Ibid.* Articles 18(1) and 28(1).

167   Joined Cases T-74/00, T-76/00 & T-141/00 *Artegodan and Others v European Commission* [2002] ECR II-4945.

168   Regulation 178/2002 laying down the general principles and requirements of food law, establishing the European Food Safety Authority and laying down procedures in matters of food safety [2002] OJ L31/1, Article 7.

adverse effects on human health, animal health or the environment'.[169] A literal interpretation of this provision would appear to endorse a zero risk approach to the 'precautionary principle', an approach that has been rejected by both the European Commission and the Court of First Instance. Equally, it is difficult to square this wording with agriculture, where most types of conventional, non-GMO, agriculture does have some adverse environmental effect.[170] Lee has suggested two possible alterative interpretations. On the one hand, she suggests that, through introducing considerations of proportionality, it could be interpreted to mean 'no unacceptable adverse effect' or 'no unacceptable risk of adverse effect'.[171] On the other hand, she suggests that GMOs might be judged according to whether any adverse effects could be identified in comparison to conventional comparators.[172] However, as Lee accepts, this latter approach involves significant assumptions and uncertainties about both GMOs and also conventional agriculture.[173]

The 'precautionary principle' is also evident in relation to measures that apply after authorisations have been granted under either the Deliberate Release Directive or the GM Food and Feed Regulation. In each case an authorisation is only valid for ten years, thus enabling new scientific evidence to be considered in deciding whether to grant its renewal.[174] Additionally, both measures require authorisation holders to act upon any new information concerning risks posed by GMOs to human health and the environment. In this situation the Deliberate Release Directive requires authorisation holders to take necessary measures to protect human health and the environment and to inform their national authority of those measures. Similarly, the GM Food and Feed Regulation requires authorisation holders to immediately inform the European Commission of any new scientific evidence or technical information which might influence the evaluation of safety in use of food. It also requires them to notify the Commission of any prohibition or restriction imposed by authorities in non-Member States. In both cases, the Deliberate Release Directive and GM Food and Feed Regulation then put in place a chain of events that may lead to the authorisation being amended or revoked.

European Community law also puts in place safeguard measures to enable individual Member States to prevent the use of approved GMOs and products containing GMOs. However, the scope to adopt such measures is tightly worded, so that, in practice, they will rarely apply. Under the Deliberate Release Directive they could only be relied upon if new scientific evidence revealed that the product concerned constitutes a risk to human health or the environment.[175] In the case of authorisations under the GM Food and Feed Regulation, national discretion is

---

169   *Ibid.* Article 4(1)(a).
170   See Lee M., footnote 107 above, 247.
171   *Ibid.*
172   *Ibid.* 248.
173   *Ibid.*
174   See Directive 2001/18, Article 15(4) and Regulation 1829/2003, Article 7(5).
175   Directive 2001/18, Article 23.

even more limited, with such safeguard measures only available either in response to an EFSA report or when it is 'evident' that the product is 'likely to constitute a serious risk to human health, animal health or the environment'.[176] The GM Food and Feed Regulation adopted an even more restrictive approach, only authorising national safeguard measures on the basis either of an opinion issued by the EFSA or that it is 'evident' that authorised products 'are likely to constitute a serious risk to human health, animal health or the environment' and, consequently, 'the need to suspend or modify, urgently, an authorisation arises'.

The 'precautionary principle' clearly has an important role to play in risk management decisions concerning the environmental or human health effects of particular GMOs. However, as noted previously, public concern about GMOs is not merely driven by these factors. Ethical and social considerations have also influenced public opinion. These broader issues may also play a significant part in the more political decision-making process. This is clearly evident in the 1998–2004 moratorium, when no new GMO applications could be granted, whatever their risk assessment. For its part, the GM Food and Feed Regulation provides that the European Commission, in drafting a decision, for submission to the Standing Committee on the Food Chain and Animal Health, on any application for authorisation under that Regulation, must take account not just of the opinion of the EFSA and of relevant provisions of Community law but also of 'other legitimate factors relevant to the matter under consideration'.[177] In practice, the scope of the phrase 'legitimate factors' may actually be quite limited. Under the Deliberate Release Directive, the protection of human health and protection of the environment were the only objectives specifically protected.[178] In contrast, the GM Food and Feed Regulation aims to achieve the broader objectives of 'ensuring a high level of protection of human life, and health, animal health and welfare, environmental and consumer interests ...'[179] However, even these wider objectives provide little scope for broader public concerns to be taken into account. For example, the Regulation distils 'consumer interests' down to the fact that consumers must not be misled and food based upon GMOs must not be nutritionally disadvantageous.[180]

The Co-existence of GM and Conventional Crops

Gmos also raise problems for agriculture in relation to the co-existence of GM crops with conventional or even organic crops. Where this happens, the risk of contamination to conventional crops by the GM crop is very real. This may result from any number of causes, such as cross-pollination by wind or insects or through

---

176   Regulation 1829/2003, Article 34.
177   Regulation 1829/2003, Article 7.
178   Directive 2001/18, Articles 1 and 4(1).
179   Regulation 1829/2003, Article 1(a).
180   *Ibid.* Article 4(1).

the use of set of farm machinery on both GM and conventional crops. As noted above, the GM Food and Feed Regulation requires food producers to use labels to advise consumers that products contain GMOs or that they have been produced from GMOs.[181] However, these requirements do not apply where no more than 0.9 per cent of a food, or of an individual food ingredient, consists of GMOs or has been produced from GMOs and the GMO presence is 'adventitious' or 'technically unavoidable.'[182] These provisions appear to provide a tolerance level for cross-contamination where GM crops co-exist with conventional crops.[183]

In order to rely upon the argument that the presence of *de minimis* GMO materials was adventitious or technically unavoidable, the GM Food and Feed Regulation provides that 'operators must be in a position to supply evidence to satisfy the competent authorities that they have taken appropriate steps to avoid the presence of such materials.'[184] In the context of co-existence this would seem to require farmers to show that they had complied with all relevant legal standards in order to avoid the occurrence of cross-contamination. This is an area that the European Community has, effectively, left to Member States. The Deliberate Release Directive provides that 'Member States may take appropriate measures to avoid the unintended presence of GMOs in other products'.[185] The European Commission has published a recommendation to Member States on national best practice concerning the co-existence of GM crops with conventional and organic farming.[186] Although not legally binding, this document has played an important role in shaping national measures.[187]The recommendation states that national measures should be efficient and cost-effective and that they should not go beyond what is necessary in order to ensure that adventitious traces of GMOs stay below the tolerance thresholds set out in Community law.[188] The recommendation sets out a list of potential measures that Member States can adopt to limit the contamination of conventional crops by GM crops, such as implementing minimum separation distances between conventional and GM crops or using separate machinery on each crop type. In the United Kingdom, measures on co-existence have yet to be agreed. However, a DEFRA consultation on measures for England suggested that

181   *Ibid.* Articles 12–14.

182   *Ibid.* Article 12(2).

183   For a critical analysis of this position see Lee, M., The Governance of Co-Existence Between GMOs and Other Forms of Agriculture: A Purely Economic Issue? [2008] 20 *Journal of Environmental Law* 193.

184   Regulation 1829/ 2003, Article 12(3).

185   Directive 2001/18, Article 26a as inserted by Regulation 1829/2003.

186   European Commission Recommendation 2003/556 *Establishing Guidelines for the Development of National Strategies and Best Practices to Ensure the Co-existence of Genetically Modified Crops with Conventional and Organic Farming*, [2003] OJ L189/36.

187   For discussion of national measures adopted in other Member States see European Commission, *Report on the implementation of national measures on the co-existence of genetically modified crops with conventional organic farming*, COM(2006) 104 final.

188   *Ibid.* para. 2.1.4.

minimum separation distances should play an important role.[189] This suggested minimum separation distances of 35 metres from GM oilseed rape crops, 80 metres from GM forage maize crops and 110 metres from GM grain maize crops.

One response to the co-existence would simply be for Member States or regional authorities to ban the use of GMOs in crops. However, it seems clear that, in practice, such a ban will be difficult to justify.[190] The European Commission's recommendation invokes the 'principle of proportionality' in providing that national measures should not go beyond what is necessary in order to ensure that adventitious traces of GMOs stay below the tolerance thresholds examined above. This would suggest that Member States must allow co-existence and that they should, instead, focus their efforts on ensuring that the 0.9 per cent threshold is not exceeded. This is also supported by the European Commission's expressed view that 'No form of agriculture, be it conventional, organic or agriculture using GMOs, should be excluded in the European Union'.[191] The European Court of Justice considered the legality of a regional ban on GMO based agriculture in *Land Oberosterrich and Republic of Austria v European Commission.*[192] Austria, here sought to rely upon Article 95(5) EC to justify a ban introduced in Upper Austria. Article 95(5) EC enables Member States to seek European Commission approval to depart from Community harmonising legislation, which, in this case, was the Deliberate Release Directive. Austria, here, challenged the European Commission's rejection of its application. Article 95(5) can only be utilised if the national measures are 'based on new scientific evidence relating to the protection of the environment or the working environment on grounds of a problem specific to that Member State arising after the adoption of the harmonising measure'.[193] Ultimately, they upheld the Commission decision as Austria had not identified any new scientific evidence to justify its proposed ban.

The authorisation of co-existence does, however, raise particular problems for organic farming.[194] On the one hand, organic produce that does not meet the 0.9 per cent threshold cannot be described as being organic.[195] On the other hand, some organic certification bodies actually apply a stricter approach. For example, the Soil Association, the United Kingdom's largest organic certification body, requires farmers to adhere to a 0.1 per cent threshold for the presence of adventitious

---

189   DEFRA (2006), *Consultation on proposals for managing the co-existence of GM, conventional and organic crops* (London: DEFRA). For analysis of these proposals see Rodgers C.P., DEFRA's Co-Existence Proposals for GM Crops: A Recipe for Confrontation? [2008] 10 *Environmental Law Review*, 1.

190   See also O'Rourke, R., footnote 51 above, 186.

191   European Commission, footnote 187 above, recital 1 to the preamble.

192   Joined Cases C-439/05P and C-454/05P, [2007] ECR-I 7141. In doing so the Court upheld a similar verdict that the Court of First Instance had already reached in the case.

193   Article 95(5) of the EC Treaty.

194   See Lee M., footnote 183 above, 205.

195   See Regulation 834/2007 on organic farming, Article 23(3).

GMOs within their produce. As Lee has pointed out, the Soil Association's stance fully accords with the objective established in the Organic Farming Regulation,[196] which provides that[197]

> The aim is to have the lowest possible presence of GMOs in organic products. The existing labelling thresholds represent ceilings which are exclusively linked to the adventitious and technically unavoidable presence of GMOs.

These provisions sit very uncomfortably with the use of the 0.9 per cent threshold.[198]

They also raise legal issues in the United Kingdom, as to how courts here make use of scientific risk analysis. For, example, as Rodgers has suggested, the private law of nuisance may have a role where organic farmers suffer financial loss due to the fact that their crops exceed the 0.1 per cent threshold required for organic certification, but are within the 0.9 per cent threshold applied in relation to the labelling of organic produce.[199] Co-existence has already raised legal issues in the United Kingdom in relation to organic farming. In *R v Secretary of State for the Environment and Minister for Agriculture, Fisheries and Food, ex parte Watson*, a farmer producing organic vegetables, challenged a decision to authorise the trial planting of GM crops on land adjacent to his farm.[200] The farmer was certified by the Soil Association, so the 0.1 per cent threshold was an important consideration. In this case, the Court of Appeal accepted scientific advice from the Advisory Committee on Releases to the Environment that, if GM crops and non GM crops were grown 200 metres apart, cross pollination would be unlikely to exceed 1 kernel in 40,000. In dismissing the case, the Court found this to be a 'perfectly reasonable point at which to strike the balance between the competing interests in play.'[201] However, if GM agriculture should become more widely adopted in future, these competing interests may come to involve the future of the organic sector as a whole.

---

196   See Lee M., footnote 183 above 207.

197   Regulation 834/2007, recital 10 to the preamble.

198   See Lee M., footnote 183, 208.

199   Rodgers, C.P., footnote 189 above 6.

200   [1999] Env L.R. 310. See also Cardwell, M., The Release of Genetically Modified Organisms into the Environment: Public Concerns and Regulatory Responses, [2002] 4 *Environmental Law Review* 156 and Campbell, D., Of Coase and Corn: A (Sort of) Defence of Private Nuisance, [2000] 63 *Modern Law Review* 97.

201   Per Simon Brown LJ.

Food Safety and the Common Agricultural Policy

As the previous paragraph observed, the Treaty objectives for the Common Agricultural Policy still make no express reference to food safety. However, it is now a Treaty requirement that all Community policies should provide a high level of health protection and also contribute towards a high level of consumer protection.[202] In the past, food safety and food quality were not central issues within the CAP. This was acknowledged by the European Commission, in putting forward its initial proposals for the 2002 Mid Term Review of the CAP. It noted that:[203]

> However, policy instruments available to support food safety and quality within the common agricultural policy remain limited. Incentives and signals sent to farmers have to be in line with the objectives of safety and quality, as they have to be in line with environmental and animal health and welfare requirements. There is a broad consensus that more can be done within the policy to meet these objectives.

In referring to food quality, these comments go beyond the scope of food safety. In this regard, the European Commission noted that the promotion of food quality was a new direction for agriculture, with consumers increasingly attracted to higher quality production and also increasingly concerned about the conditions in which agricultural products were produced.[204] Equally, though, the Commission accepted that 'there remains a gap between the preference for quality that consumers express and the way that they behave in market place'.[205]

Issues of food safety and food quality have become a more integral part of the Common Agricultural Policy since the Mid Term Review. Regulation 1872/2003 introduced cross-compliance obligations concerning food safety.[206] To remain eligible to receive direct payments, under agricultural production policy, farmers must currently comply with the following European Union food safety laws:

- Directive 91/414/EEC concerning the placing of plant protection products on the market[207]
- Directive 96/22/EEC concerning the prohibition of the use in stock farming of certain substances having a hormonal or thyrostatic action and of beta-agonists[208]

---

202   See Articles 152 and 153 respectively.
203   European Commission, *Mid Term Review of the Common Agricultural Policy*, COM (2002)394, 7.
204   *Ibid.* 7.
205   *Ibid.* 6.
206   [2003] OJ L270/1.
207   [1991] OJ L230/1.
208   [1996] OJ L125/3.

- Regulation 178/2002 of the European Parliament and of the Council laying down the general principles and requirements of food law, establishing the European Food Safety Agency and laying down procedures in matters of food safety[209]
- Regulation 999/2001 of the European Parliament and of the Council laying down rules for the prevention, control and eradication of certain transmissible spongiform encephalopathies[210]
- Directive 85/511/EEC introducing Community measures on the control of foot and mouth disease[211]
- Directive 92/119/EEC introducing general Community measures for the control of certain animal diseases and specific measures relating to swine vesicular disease[212]
- Directive 2000/75/EC laying down specific provisions for the control and eradication of bluetongue, OJ [2000] L327/74[213]

Additionally, the European Community's rural development policy now enables Member States to provide financial support for measures that support food safety or promote food quality. Under Axis 1 of rural development policy Member States are authorized to assist farmers, on the basis of costs incurred and income foregone, in meeting new standards in environmental protection, public health, animal and plant health, animal welfare or occupational safety.[214] Similarly, they can also provide support to farmers participating in schemes to enhance food quality as well as supporting producer groups to promote food quality measures.[215]

---

209 [2002] OJ L31/1.
210 [2001] OJ L147/1.
211 [1985] OJ L315/11.
212 [1992] OJ L62/69.
213 [2000] OJ L327/74.
214 Council Regulation 1698/2005, [2005] OJ L277/1, Article 31.
215 *Ibid.* Articles 32 and 33.

Chapter 10

Agriculture, the Environment and International Trade

Regulating International Trade

Beyond the borders of the European Community, trade in agricultural commodities is also an important aspect of international trade. The international community has played an active role in seeking to facilitate and encourage international trade. In the aftermath of the Second World War, the General Agreement on Tariffs and Trade (GATT) was signed in 1947, with the objective of establishing a framework for agreement on international trade and a means by which protectionist measures could be discouraged.[1] GATT also sought to encourage trade liberalisation by requiring contracting states to, periodically, review the Agreement, through, so-called, 'trade rounds'. In 1994 the eighth such round, the Uruguay Round, agreed to disband the GATT and to replace it with the World Trade Organisation (the WTO).[2]

The developing framework of international trade law has had an important influence upon the Common Agricultural Policy and on the way environmental protection issues have been integrated within it. This is particularly evident in the 1994 WTO Agriculture Agreement, which restricted the European Community's freedom designing CAP policy measures.

Although the international community has sought to liberalise international trade, by removing protectionist measures, it has also recognised that nations do have legitimate concerns that may justify restrictions on trade. In this area, environmental issues, whether they relate to a desire to keep out invasive species or to prevent animal or plant diseases from spreading to their territories, provide an obvious example of such concerns. However, this also creates the danger that countries may seek to use such provisions to justify measures that discriminate in favour of domestic produce. International trade law has, therefore, had to balance

---

1   For an examination of the role of GATT and the WTO in relation to world trade in agricultural produce see generally Josling, T., Tangermann, S., Warley, S. (1994), *Agriculture in the GATT* (London: Macmillan) chapter 1 – The GATT's Origins and Early Years or McMahon, J.A., From Havana to Seattle: A History of Trade and Agriculture, in McMahon, J.A. (ed.) (2001), *Trade and Agriculture Negotiating a New Agreement?* (London: Cameron May).

2   For a detailed analysis of the WTO see, Van den Bossche, P. (2005), *The Law and Policy of the World Trade Organisation* (Cambridge: Cambridge University Press).

the interest of environmental protection against the desire to ensure that countries do not simply use environmental measures as a cloak for protectionist measures.

International Trade Reform and the Common Agricultural Policy

The formation of GATT, in 1947, provided a mechanism to facilitate the liberalisation of international trade. However, trade in agricultural produce was generally excluded from the GATT Agreement.[3] This had the result that, whilst the trade rules in other sectors were gradually liberalised, those dealing with agricultural produce provided increasing protection for national farming industries.[4]

Agriculture's exclusion from international trade discussions effectively remained in place until 1986. That year marked the opening of the Uruguay Round of negotiations on trade liberalisation. In particular, the United States and a group of developed agricultural exporting nations collectively known as the 'Cairns Group' made it clear that if the liberalisation of national farming policies was not included they would not support other aspects of these trade discussions.[5] It has been estimated that the Cairns group of nations collectively sustained a 50 per cent reduction in the value of their food exports as a result of measures taken to subsidise agricultural production in other countries.[6] The United States, similarly, sought to arrest a decline in its agricultural exports.[7]

---

3   In reality the provisions of GATT applied equally to trade in agricultural and industrial products. However, a number of special exemptions applied in the case of agriculture. For example, whilst Article XI (i) of GATT prohibited the use of qualitative import restrictions in international trade, Article XI (ii) allowed such restrictions to remain for agricultural produce where they were necessary to enforce domestic market management. Similarly Article XVI(iv) prohibited the payment of export subsidies, but Article XVI(iii) provided that they could be paid in relation to agricultural produce as long as they did not result in any nation gaining more than an 'equitable share' in world trade in a particular product.

4   See Harvey, D., The GATT, the WTO and the CAP, in Ritson C., Harvey, D.R. (eds) (1997), *The Common Agricultural Policy*, 2nd ed. (Wallingford: CAB International) 378 and Greenway, D., The Uruguay Round: Expectations and Outcomes, in Ingerset, K.A., Rayner, A.J., Hine, R.C. (eds) (1994), *Agriculture in the Uruguay Round* (London: Macmillan) 12.

5   *Ibid.* 379. The Cairns group comprised: Argentina, Australia, Brazil, Canada, Chile, Columbia, Fiji, Hungary, Indonesia, Malaysia, New Zealand, the Philippines, Thailand and Uruguay and was named after the location of a conference that they held before the commencement of the Uruguay Round.

6   Tyers, R. The Cairns Group Perspective, in Ingerset, K.A., Rayner, A.J., Hine, R.C. (eds), *Agriculture in the Uruguay Round*, footnote 4 above, 97.

7   For example, Josling has noted that in the period between the mid 1970s and 1987 the European Community first became self-sufficient in wheat production and then moved on to capture 15 per cent of the world wheat market. In contrast the United States share of the world wheat market fell from 48 per cent in 1981 to 30 per cent in 1985. See, Josling, T.,

The measures adopted within the CAP were a particular target for these countries. The principle of 'Community preference' together with the application of variable import levies restricted accessibility to European Community markets for imported products. Simultaneously, the intervention system insulated farmers from fluctuations in world agricultural prices and provided an artificial market for their produce. Additionally, the payment of export refunds allowed farmers to compete upon world markets without sustaining income losses. As if that were not enough, the European Community had evolved from being a net importer to being a net exporter of agricultural produce. Across the globe, agricultural exporting nations were also experiencing increased agricultural productivity. The net result was that world export markets were becoming increasingly competitive and domestic agricultural support was becoming increasingly expensive for competing nations.

The Cairns Group and the United States sought radical reductions in levels of export refunds and domestic agricultural support within the Common Agricultural Policy, which they perceived as giving the European Community an unfair advantage in world export markets.[8] In addition, they sought greater access to European Community markets, through the removal of variable import levies and limitations on the use of the principle of 'Community preference'.[9] Such measures threatened to undermine the operation of the Common Agricultural Policy. Imported produce would be able to undercut the price of European Community produce, whilst large reductions in export refund and intervention payments would lead to sharp falls in agricultural incomes across the Member States. Alternatively, failure to reach an agreement on agricultural trade reform created the danger of a trade war that excluded the European Community's agricultural and industrial products from world markets and damaged Member States' economies.

Ultimately, agricultural trade issues played a central role in the Uruguay Round negotiations. Indeed, agreement on broader international trade reforms was delayed, whilst negotiations continued on the liberalisation of agricultural trade. The measures agreed upon were set out in the Agriculture Agreement, a stand alone document dealing solely with agricultural trade. The preamble to the Agreement sets the objective of establishing a 'fair and market-orientated agricultural

---

The CAP and North America, in Ritson, C.R., and Harvey, D.R., *The Common Agricultural Policy*, footnote 4 above, 363.

8   See Potter C., Ervin, D.E., Freedom to Farm: Agricultural Policy Liberalisation in the US and EU, in Redclift, M.R., Lekakis, J.N., Zanais, G. (eds) (1999), *Agriculture and World Trade Liberalisation: Socio-Environmental Perspectives on the Common Agricultural Policy* (Wallingford: CAB International) 55.

9   See Hillman, J.S., The US Perspective and Tyers, R., The Cairns Group Perspective, both in Ingerset, K.A., Rayner A.J., and Hine, R.C. (eds), *Agriculture in the Uruguay Round*, footnote 4 above. See also Josling, T., Tangermann, S., Warley, T. (eds) (1996), *Agriculture in the GATT* (London: Macmillian) at chapter 7, 'The Uruguay Round'.

trading system'.[10] This was to be achieved by 'substantial progressive reductions in agricultural support and protection', 'resulting in correcting and preventing restrictions and distortions in world agricultural markets.'[11] The Agreement itself then established binding commitments for member countries in three principal areas: market access, export competition and domestic support.[12]

Market Access

To facilitate exports, member countries were required to undertake a process of tariffication, replacing non-tariff barriers to trade in agricultural produce with customs tariffs.[13] They were then required to gradually reduce these tariffs over the period 1995 to 2000. Developed countries, along with the European Community, gave a commitment to reduce their tariffs by an average of 36 per cent over this period, from a base tariff level that had existed in the period 1986–1988.[14] In contrast, developing countries committed themselves to an average 24 per cent reduction over ten years.

Export Competition

The Agriculture Agreement also included a commitment, by member countries, to reduce the export subsidies they paid to producers. Developed countries agreed to reduce subsidised exports by 36 per cent in value and by 21 per cent in volume between 1995 and 2000, from levels experienced in a base period between 1986 and 1990. Developing countries gave a commitment to make corresponding reductions of 24 per cent and 14 per cent over a ten-year period.

---

10   Agriculture Agreement 1994, para. 2 to the preamble.

11   *Ibid.* para. 3.

12   See generally Cardwell, M., Rodgers, C., Reforming the WTO legal order for agricultural trade: issues for European rural policy in the Doha Round, [2006] 55 *International Comparative Law Quarterly* or McMahon, J.A. (2000), *Law of the Common Agricultural Policy* (London: Pearson Education) 151–158.

13   Articles 4 and 5 and Annex 5.

14   In practice this gave rise to the practice of 'dirty tariffication', in which countries chose a base period in which their tariffs were high, thus ensuring that relatively high tariffs could remain in place even after the reductions required by the Agriculture Agreement, see Harvey, D., The GATT, the WTO and the CAP, in Ritson C., Harvey, D.R. (1997), *The Common Agricultural Policy* (Wallingford: CAB International) 385 or Josling, T., Tangermann, S., and Warley, T. (eds) (1996), *Agriculture in the GATT* (London: Macmillan) chapter 7.

Domestic Support

Member countries agreed to reduce the amount of domestic support that they provided to producers.[15] Their reduction commitment is expressed in terms of 'aggregate measurement of support', which is defined as being[16]

> ...the annual level of support, expressed in monetary terms, provided for an agricultural product in favour of producers of the basic agricultural product or non product specific support provided in favour of agricultural producers in general.

These payments are generally referred to as being 'amber box' payments. They are regarded as being payments that encourage production and help to distort international trade. Developed countries agreed to reduce such payments by 20 per cent over the period 1995 to 2000, whilst developing countries agreed to achieve a 13 per cent reduction over 10 years and least developing countries were required to observe a moratorium, under which payments could not be increased. However, the Agriculture Agreement also exempted a range of other domestic support payments from this reduction requirement. The exempt payments were either regarded as being of a *de minimis* level or as being 'green box' or 'blue box' payments.

Domestic support for the production of specific produce was considered to be *de minimis* if it did not exceed 5 per cent of the total value of a member's production of that particular produce in any year.[17] Additionally, non product specific domestic support was considered to be *de minimis* if it did not exceed 5 per cent of the value of a members total agricultural production.[18] These percentages were increased to 10 per cent for developing-country members.[19] 'Green box' payments were also excluded from the requirement to reduce domestic support. Annex 2 to the Agreement lists 12 specific payment types that fall into this category. These include payments to producers that are decoupled from production, payments made to producers enrolled in environmental programmes and payments made to producers in disadvantaged areas. In each case, in addition to further policy specific criteria, the exemption will only apply if the payments satisfy the following conditions:[20]

> Domestic support payments for which exemption from the reduction requirements is claimed shall meet the fundamental requirement that they have no, or at

---

15   Agriculture Agreement, Article 6.

16   Agriculture Agreement, Article 1(a). See also Annex 3 to the Agreement, which makes detailed provision for the calculation of the aggregate measurement of support.

17   Agriculture Agreement, Article 6(4) (a).

18   *Ibid.*

19   Agriculture Agreement, Article 6(4) (b).

20   Agriculture Agreement, Annex 2, para. 1.

most minimal, trade distortion effects or effects on production. Accordingly all policies for which exemption is claimed shall conform to the following general criteria:

a. the support in question shall be provided through a publicly-funded government programme (including governmental revenue foregone) not involving transfers from consumers
b. the support in question shall not have the effect of providing price support to producers

Additionally, the recognition of 'blue box' payments was based upon the notion that these payments caused only minimal trade distortions. This resulted from a political compromise between the United States and the European Community, which have been the principal beneficiaries of its creation.[21] The compromise ensured that members would not be required to reduce direct payments made to producers under production-limiting programmes provided those payments met one of three conditions:[22]

a. they were based upon fixed areas and yields
b. they were made on 85 per cent or less of the base area level of production
c. they were livestock payments that were made for a fixed number of head

Finally, the Agriculture Agreement also provided for WTO members to commence further agricultural trade negotiations in 2000, with the aim of taking further steps towards the 'long term objective of substantial progressive reductions in support and protection...'.[23] These negotiations commenced in March 2000 and, at the time of writing, remained ongoing.[24] Under the terms of the Agriculture Agreement the exemptions provided, to de minimis payments and to both blue and 'green box' payments, were granted for a nine-year period 1995 to 2003.[25] The continued existence of these payments has been a central issue in the current agricultural trade negotiations.

As might be expected, the Agriculture Agreement and the current negotiations on further liberalisation of agricultural trade have had a profound impact on the CAP. Both the MacSharry and Agenda 2000 reforms were heavily influenced

---

21 Grossman, M.R., The URAA and Domestic Support, in Cardwell, M., Grossman, M.R. and. Rodgers, C.P., (eds) (2003), *Agriculture and International Trade; Law, Policy and the WTO* (Wallingford: CAB Publishing) 40.

22 Agriculture Agreement, Article 6(5).

23 Agriculture Agreement, Article 20.

24 As had been the case in the previous Uruguay Round, agreement on the liberalisation of broader international trade issues has been stalled by failure to reach agreement on agricultural trade reforms.

25 Agriculture Agreement, Article 13.

by the negotiations leading up to the 1994 Agriculture Agreement and by the Agreement itself, whilst the 2003 Mid Term Review and 2008 CAP Health Check reforms sought to strengthen the European Community's position in anticipation of a future WTO agreement.

In response to the Agriculture Agreement, the European Community replaced the variable levies, previously charged upon imported produce, with fixed customs duties.[26] Other commentators have noted that the main problem that the European Community faced, in meeting its Agriculture Agreement commitments, was ensuring that it achieved sufficient volume reductions in subsidised exports.[27] As noted in Chapter 1, the MacSharry and Agenda 2000 reforms resulted in reductions to prices, set by the European Community for commodities such as arable crops, milk and livestock, that were intended to reduce market prices within the European Community. This reduced the gap between European Community market prices and the lower prices historically prevailing on world markets. This meant that there would be less need for export refunds to be paid to compensate producers for this difference. Additionally, some produce, for example sheep and goats meat, became ineligible for export refund payments,[28] whilst payments were only available for other produce in so far as the European Community did not exceed its Agriculture Agreement obligations.[29] Elsewhere, the range of direct payments introduced to compensate farmers for price reductions, and the more limited availability of intervention funding, were linked to production limiting requirements. For example, arable area payments could only be made in respect of land that was being used for rotational crops on 31 December 1991, whilst the set-aside requirements placed further limits on production.[30] Consequently, they provided no incentive for farmers to bring new land into cultivation. In the case of livestock payments, payments such as suckler cow premium[31] and sheep annual premium[32] were based upon historic livestock numbers. Applicants for suckler cow premium also had to meet maximum livestock density levels. Collectively,

---

26   See Regulation 3290/94, [1994] Official Journal L349/105.

27   See Harvey, D., The GATT, the WTO and the CAP, in Ritson C., and Harvey D.R., (eds) (1997), *The Common Agricultural Policy* (Wallingford: CAB International) 387.

28   See Regulation 3013/89 Title 2 ([1989] Official Journal L289/1) as inserted by Regulation 3290/94 ([1994] Official Journal L349/105).

29   For arable crops see Regulation 1766/92 ([1992] Official Journal L181/21) as inserted by Regulation 3290/94 (Official Journal L349/105), for milk products see Regulation 1255/99, Article 31 ([1999] Official Journal L160/48), for beef and veal see Regulation 1254/99, Article 33 ([1999] Official Journal L160/21). In each case the volume of subsidised exports is calculated by reference to the number of export certificates issued.

30   As originally introduced by Regulation 1765/92, [1992] Official Journal L181/12.

31   By virtue of Council Regulation 2066/92, [1992] OJ L215/49. See also Neville, W., Mordaunt F., (1993), *A Guide to the Reformed Common Agricultural Policy* (London: Estates Gazette) 84.

32   By virtue of Council Regulation 2069/92, [1992] OJ L215/59. See also Neville, W., Mourdant, F., footnote 31 above, 115.

these provisions removed incentives for farmers to increase stocking densities. By discouraging increased production such requirements helped to address the production surplus problems that the European Community was experiencing and, by reducing the levels of produce available for export, limiting the need to pay export refunds. The payments themselves were linked to production requirements, since arable and livestock farmers needed to continue in production, within their particular sector, in order to remain eligible. As such, none of the payments qualified for inclusion within the 'green box', however, they were subject to production limiting requirements and were, therefore, accepted as being 'blue box' payments. Consequently, they were not subject to the European Community's obligation to reduce domestic support payments.

The more recent 2003 Mid Term Review of the CAP and the 2008 Health Check were both reform packages introduced with an eye to the on-going negotiations on the liberalisation of agricultural trade. In particular, the Mid Term Review sought to address the position of CAP's direct support payments within the 'blue box'. Basically, the 'blue box' had been the result of a politically-expedient agreement when the 1994 Agriculture Agreement had been made. Its existence, primarily, benefited the farm policies of the European Community and United States and there was likely to be considerable pressure from other members for it to be abolished. Consequently, the European Community has attempted to adapt its current direct payment, the Single Farm Payment, to bring it within the more easily defensible 'green box'. Under Annex 2 of the Agriculture Agreement, only income support payments that make no production requirements of producers, can be accepted as coming within the 'green box'. As noted in Chapter 1, the European Commission's proposals for the Mid Term Review sought to address this problem by suggesting that the Single Farm Payment should become entirely decoupled from production. However, the amendments accepted by the Council, enabling Member States to continue to link eligibility for a portion of the Single Farm Payment to production-based requirements, means that the Single Farm Payment still falls short of 'green box' acceptability. Additionally, as Rodgers has pointed out, the introduction of compulsory cross-compliance under the Mid Term Review could, rather ironically, pose similar problems.[33] The fact that producers must now maintain their land in 'good agricultural and environmental condition' would necessitate continued production and, therefore, could be considered to be a production requirement that falls fowl of the current 'green box' requirements.

Elsewhere, the major elements of the European Community's rural development policy, agri-environmental land management and less favoured area payments, could also be considered to be 'green box' payments and also exempt from the obligation to reduce domestic support. However, this is not beyond dispute. For example, Rodgers has noted that the agri-environmental land management schemes, introduced by Member States, have generally pursued objectives with a wider

---

33   Rodgers, C.P., Renegotiating the WTO Agreement on Agriculture: Biodiversity and Agricultural Policy Reform at a Crossroads? [2004] 6 *Environmental Law Review* 73.

scope than simply promoting environmentally-friendly farming.[34] The Agriculture Agreement provides that 'payments under environmental programmes' form part of the 'green box'. However, no definition of 'environmental programmes' is provided. It is not clear that schemes with environmental and other, wider, objectives will be accepted as being truly 'green box' compliant. Equally, there is also the fact that the Agriculture Agreement provides that payments under environmental programmes must be limited to the extra costs or loss of income involved in complying with the terms of the programme.[35] Previously, the European Community had authorised Member States to base agri-environmental land management payments on income foregone, additional costs incurred by participating producers and the need to provide an incentive to encourage participation.[36] The incentive could be up to 20 per cent of the income foregone and extra costs incurred by the producer.[37] The payment of such incentive payments could have jeopardised the status of these payments within the 'green box'.[38] The European Community moved to address this latter problem in its 2005 Mid Term Review reforms. Under Regulation 1698/2005 payments are, henceforth, to be calculated solely on the basis of the income foregone by farmers as a result of taking part in the scheme and also any additional costs that might arise from their participation.[39] However, as examined in Chapter 5, this could impact on the success of these agri-environment measures. Questions could, equally, be posed as to the 'green box' compatibility of the European Community's less favoured area payments. As was noted in Chapter 4, the Court of Auditors, in 2003, noted that over 56 per cent of agricultural land in the, then, 15 Member States, had been identified as being agriculturally less favourable.[40] They raised concerns about the fact that land had been identified as being less favourable on the basis of outdated socio-economic data and questioned the validity of these identifications.[41] These observations sit uncomfortably with the Agriculture Agreement's requirement that payments under regional assistance programmes should only qualify for inclusion, within the 'green box', if those regions are identified on 'the basis of neutral and objective criteria clearly spelt out in law or regulation and indicating that the region's difficulties arise out of more than temporary circumstances.'[42] The renewed international trade negotiations, provided for in Article 20 of the Agriculture Agreement, commenced in March

---

34   *Ibid.* 75.

35   Agriculture Agreement, Annex 2 para. 12(b).

36   See for example Regulation 1257/99, Article 24, [1999] Official Journal L160/80.

37   See Commission Regulation 445/2002, Article 19, [2002] Official Journal L74/1.

38   Rodgers, C.P., footnote 33 above, 74.

39   Article 39(4). See also Commission Regulation 1974/2006 laying down detailed rules for the application of Regulation 1698/2006 ([2006] OJ L368/1) Article 53.

40   Court of Auditors, Special Report 4/2003 *Concerning Rural Development: Support for Less Favoured Areas*, [2003] OJ L151/12, 14.

41   *Ibid.*

42   Agriculture Agreement, Annex 2 para. 13(a).

2000. Agricultural trade is only one of a range of trade issues that are being discussed within these negotiations. In November 2001 the WTO Ministerial Council, meeting at Doha in Qatar, issued a declaration explaining their objectives and providing a timetable for the negotiations.

Under Article 13 of this declaration, the so-called Doha Declaration, the WTO members committed themselves to 'comprehensive negotiations aimed at: substantial improvements in market access; reductions of, with a view to phasing out, all forms of export subsidies; substantial reductions in trade-distorting domestic support.'[43] These mirror the principal trade issues which framed the negotiations that led to the 1994 Agriculture Agreement. Key questions have been whether 'amber box' payments should be substantially reduced or eliminated altogether and, similarly, whether 'blue box' payments should be jettisoned or allowed to continue. Additionally, the role of non-trade concerns, within WTO member's agricultural policies, has been another important issue.

The issue of non-trade concerns basically recognises that agriculture can produce a wide range of public goods as well as agricultural produce. Paragraph six of the preamble to the 1994 Agriculture Agreement refers to two such non-trade concerns – food security and the need to protect the environment. However, the question that WTO members have had to address is whether, and to what extent, national agricultural policies should provide financial support for such objectives. In the case of environmental protection, agriculture has both negative and positive environmental impacts. The 'polluter pays' principle might dictate that producers themselves should bear the cost of tackling agricultural pollution issues. But should agricultural policies provide financial payments to producers who take positive action, such as creating or maintaining wildlife habitats on their land? In the absence of these payments, no other 'environmental market' exists to encourage such action. Indeed, agricultural policy would then actively discourage it, since farmers would be better off, economically, were they to ignore wildlife issues and simply use the land to its full agricultural potential.

The role of non-trade concerns was originally recognised in Article 20 of the Agriculture Agreement, which required WTO members to take these non-trade concerns into account as part of the renewed negotiations. Similarly, Article 13 of the Doha Declaration also confirms that 'non-trade concerns will be taken into account in the negotiations'. In practice, as one commentator has noted, the question of whether the WTO Agriculture Agreement should regulate measures designed to promote non-trade concerns, and if so how, has been a major obstacle within the agriculture negotiations.[44]

In the case of the European Community, non-trade concerns play a central role within the Common Agricultural Policy. The European Community views agriculture

---

43   Doha Declaration, Article 13, available at http://www.wto.org.

44   Smith, F., Non Trade Concerns and Agriculture in a Post-Doha Environment: Thinking Outside the Green Box, [2007] 9 *Environmental Law Review* 90, see also Cardwell, M., Rodgers, C., footnote 12 above.

as being a 'multifunctional' policy and supports a range of non-trade concerns such as the environment and biodiversity, supporting rural communities, improving food quality and improving animal welfare. These are supported through financial resources available through Rural Development policy and compliance with basic obligations in these areas, with the exception of support for rural communities, has become a cross-compliance requirement for Single Farm and Less Favoured Area payments. Consequently, the European Community has argued strongly that national agricultural policies should be able to continue to support such objectives.[45]

However, the European Community's position is not shared by all WTO members. In particular, the Cairns group has been strongly opposed to the concept of non-trade concerns. Their opposition relates to two issues. Firstly, that the argument about the need to support multifunctional objectives within national agricultural policies is being used by countries that traditionally followed highly protectionist agricultural trade policies as a 'disguise' behind which they can continue to protect their agriculture sectors.[46] Secondly, as payments in relation to non-trade concerns can operate to distort trade, the 'green box' should be amended to ensure that only policies that had no, or minimal, trade-distorting effects could be adopted.[47] It is certainly true that such payments can have impacts on agricultural trade. By increasing incomes they can, indirectly, encourage producers to keep land in agricultural use or take other production decisions that increase productivity and, therefore, competition for imported produce. This, in a sense, highlights the paradox between agricultural policy and the promotion of non-trade concerns, since the best measure to employ to address the non-trade concern may not necessarily be the least trade-distorting.[48] The European Community asserts that the measures most suited to achieving the non-trade concern goal should be adopted not simply those that have the least impact upon trade.[49] For their part, the United States adopted a position somewhere between these two viewpoints – arguing that national programmes should ensure that any support payments are fully decoupled from production and directly targeted to the particular non-trade

---

45   For a discussion of the European Community's negotiating position in the trade negotiations see Cardwell, M., 2004. *The European Model of Agriculture* (Oxford: Oxford University Press) 333–341, Grossman, M.R., Multifunctionality and Non-trade Concerns and Cardwell, M., Multifunctionality of Agriculture: A European Community Perspective, both in Cardwell, M., Grossman M.R., and Rodgers, C.P. (2003), *Agriculture and International Trade: Law, Policy and the WTO* (Wallingford: CAB International) or O'Neill, M., Agriculture, the EC and the WTO: a legal critical analysis of the concepts of sustainability and multifunctionality, [2002] *Environmental Law Review* 144.

46   See Grossman, M.R., Multifunctionality and Non-trade Concerns, in Cardwell, M. Grossman, M.R., and Rodgers, C.P. (eds), *Agriculture and International Trade: Law, Policy and the WTO*, footnote 45 above, 100–102.

47   *Ibid.*

48   F. Smith, footnote 44 above, 99.

49   *Ibid.*

concern so as to minimise any impact on agricultural trade.[50] So, for example, producers should only receive payment for creating or maintaining a specific habitat on their farm, as opposed to receiving a general payment for enrolling their farm in an environmental land management programme. Ultimately, the outcome of the current round of negotiations will have an important effect upon the future content of the Common Agricultural Policy. In particular, any agreement on the multifunctional role of agriculture, in relation to non-trade concerns, will help to determine the future shape of the 'green box' within the WTO Agriculture Agreement and strongly influence the content of the European Community's Rural Development.

World Trade and Environmental Protection

In practice, GATT always enabled countries to justify restrictions on international trade on environmental grounds. Under Article XX (b) and (g) of the 1994 GATT Agreement (and, indeed, also the preceding 1947 Agreement) contracting parties were authorised to do so on the grounds that the measures adopted were '(b) necessary to protect human, animal or plant life or health' or '(g) relating to the conservation of exhaustible natural resources if such measures are made in conjunction with restrictions on domestic production or consumption'. In either case, Article XX makes it clear that the national measures cannot be applied in a manner that would constitute an arbitrary discrimination between countries in which the same conditions prevail or as a disguised restriction on international trade.

Disputes regarding the application of Article XX (b) and (g) or indeed other questions concerning the compatibility of national measures with international trade law, will be settled through agreed dispute settlement procedures. The 1994 GATT Agreement provides a structured dispute mechanism through which to settle trade disputes between member countries.[51] Overall responsibility for dispute resolution lies with a Dispute Settlement Body, comprised of all WTO member countries. Where attempts to settle disputes through conciliation are unsuccessful then WTO dispute resolution panels will, usually, be established to rule upon the dispute, with a right of appeal, on points of law, to an Appellate Body. The reports prepared by

---

50   See Sensi S., Werksman J. Inside the Green Box: 'Multifunctional' Agriculture and the Protection of the Environment, in McMahon, J.A., (ed.) (2001), *Trade and Agriculture: Negotiating a New Agreement?* (London: Cameron May) 463.

51   See Annex 2 to the 1994 GATT Agreements, Understanding on Rules and Procedures Governing the Settlement of Disputes. For an in-depth examination of the understanding see Marceau, G., Morrissey, J., Clarification of the Dispute Settlement Understanding brought by WTO Jurisprudence in McMahon, J.A., (ed.) (2001), *Trade and Agriculture: Negotiating a New Agreement* (London: Cameron May).

these panels and Appellate Bodies play a prominent part in the binding settlement to each trade dispute adopted by the Dispute Settlement Body.

In the past, GATT panels adopted a restrictive approach to Articles XX (b) and (g).[52] As One panel noted:[53]

> ...the practice of the panels has been to interpret Article XX narrowly, to place the burden of proof on the party invoking Article XX to justify its invocation and not to examine Article XX unless invoked.

More recently, WTO panels have also continued to apply this restrictive approach. In the US Gasoline case the WTO Appellate Body accepted that clean air was an exhaustible natural resource governed by Article XX (g) and that US standards on gasoline aimed to protect it.[54] However, the fact that these standards were applied differently to domestically produced and imported gasoline resulted in unjustifiable discrimination and was a disguised restriction on trade. Similarly, in the Shrimp-Turtle case, US measures to prohibit shrimp imports caught using equipment that endangered sea turtles were also condemned. The WTO Appellate body in this case recognised that references to 'exhaustible natural resources' in Article XX (g) covered living endangered species such as the sea turtle. It was also satisfied that the US legislation was concerned with their conservation and had been adopted in conjunction with similar restrictions on domestic production. However, the Appellate Body rejected the 'rigid and unbending standard' of the US measures, which effectively required other countries to adopt the same regulatory measures as the US. This was found to constitute an arbitrary discrimination, which prevented the US measures from satisfying Article XX (g). In general both GATT and WTO panels have expressed preference for multilateral action at international level, rather than the adoption of unilateral environmental measures by one GATT/WTO member. In the Shrimp-Turtle case the WTO Appellate body noted that:[55]

> ... general international law and international environmental law clearly favour the use of negotiated instruments rather than unilateral measures when addressing transboundary or global environmental problems, particularly when developing

---

52 See DS21/R United States: Restrictions on Import of Tuna, Panel Report, 30 ILM 1594 (1991) and DS29/R United States: Restrictions on Import of Tuna, Panel Report, 33 ILM 839 (1994). For an examination of the general approach of GATT and WTO panels see Redgwell, C., Trade Measures and Environmental Protection, in McMahon, J.A., (ed.) (2001), *Trade and Agriculture: Negotiating a New Agreement?* (London: Cameron May).

53 DS 21/R United States: Restrictions on Import of Tuna, Panel Report, 30 ILM 1594 (1991) at para. 5.22.

54 WTO Appellate Body Report, *United States – Standards for Reformulated and Conventional Gasoline*, WT/DS2/AB/R.

55 *Ibid.* para. 7.61. For the expression of similar views by GATT panels see the panel decisions in Tuna-Dolphin I and II, as cited at footnote 52 above.

countries are concerned. Hence a negotiated solution is clearly to be preferred, both from WTO and an international environmental law perspective.

The WTO's Appellate Body has accepted that WTO members have the right to determine the level of protection of health that they consider appropriate. Consequently, France was found to be entitled to introduce a ban on both domestic and imported asbestos and of substances containing asbestos.[56] This ban was acceptable since it applied equally to domestic and imported produce and was based upon 'the well known and life threatening health risks posed by asbestos fibres'.[57]

Augmenting World Trade's Environmental Standards

Article XX of the GATT Agreement provided some recognition of the fact that countries may have legitimate reasons for restricting trade with third countries. Equally, it also shows that the international community has long been very much aware of the need to ensure that such measures are not abused, in order to give economic protection to national industries. As part of the Uruguay Round of negotiations, the international community put in place more elaborate measures to augment Article XX and provide further guidance on the dividing line between legitimate trade restrictions and measures that either amounted to arbitrary discrimination or disguised restrictions in international trade. In particular, three agreements were signed that have particular relevance to agricultural trade. These were the Agreement on Sanitary and Phytosanitary Measures ('the SPS Agreement'), the Agreement on Technical Barriers to Trade ('the TBT Agreement') and the Agreement on Trade Related Aspects of Intellectual Property Rights ('the TRIPs Agreement').

The Agreement on Sanitary and Phytosanitary Measures

The SPS Agreement is directly concerned with steps that countries take to adopt or enforce measures necessary to protect human, animal or plant life or health. One of its principal aims is to ensure that such measures can be harmonised, as much as possible, throughout world trade.[58] As such, the Agreement has obvious implications in regulating the ability of individual countries and trade blocks, such as the European Community, to introduce trade restricting environmental measures. Under the Agreement, sanitary and phytosanitary measures are broadly

---

56 WTO Appellate Body Report, *European Communities – Measures Affecting Asbestos and Asbestos Containing Products*, WT/DS135/AB/R.

57 *Ibid.* para. 169.

58 See the preamble to the agreement and also Article 3.

defined to include 'all laws, decrees, regulations, requirements and procedures'[59] that are applied:

a. To protect animal or plant life or health, within the territory of the Member, from risks arising from the entry, establishment or spread of pests, diseases, disease-carrying organisms or disease-causing organisms
b. To protect human or animal life or health, within the territory of the Member, from risks arising from additives, contaminants, toxins or disease-causing organisms in foods, beverages or feedstuffs
c. To protect human life or health, within the territory of the Member ,from risks arising from diseases carried by animals, plants or products thereof, or from the entry, establishment or spread of pests
d. To prevent or limit other damage, within the territory of the Member, from the entry, establishment or spread of pests.[60]

However, the WTO panel decision in *EC-Biotech Products* illustrates the potentially wider scope of the agreement.[61] The panel considered complaints by the United States, Canada and Argentina that the European Community had acquiesced whilst a number of its Member States had operated moratoria on the approval of GM products. Five Member States had made a declaration, in June 1999, stating that they intended to prevent all future applications for approval of such products until such times as the EU introduced legislation concerning the labelling and traceability of GM products.[62] This legislation was not adopted until September 2003.[63] Neither the declaration nor the evident inaction, at both national level and European Community level, in approving GM products, came within the definition of sanitary and phytosanitary measures. Nonetheless, they

---

59   Annex A, Article 1. This Article provides that such measures include, inter alia, end product criteria; processes and production methods; testing, inspection, certification and approval procedures; quarantine treatments including relevant requirements associated with the transport of animals or plants, or with the materials necessary for their survival during transport; provisions on relevant statistical methods, sampling procedures and methods of risk assessment; and packaging and labelling requirements directly related to food safety.

60   *Ibid.*

61   WT/DS291/R, WT/DS292/R and WT/DS293/R European Communities – Measures Affecting the Approval and Marketing of Biotech Products. For a detailed analysis of this case see Cheyne, I., Life after the Biotech Products Dispute, [2008] *Environmental Law Review*, 52.

62   The Member States concerned were Denmark, France, Greece, Italy and Luxembourg. They made the declaration at the Council of the European Union's 2194 meeting.

63   Regulation 1830/2003 concerning the traceability and labelling of genetically modified organisms and the traceability of food and food products from genetically modified organisms, [2003] OJ L266/24.

were still in breach of Article 8 and Annex C, paragraph 1(a) of the Agreement, which requires SPS approval mechanisms to be 'undertaken and completed without undue delay'.

The SPS Agreement asserts that WTO Members have the basic right to take sanitary and phytosanitary measures necessary to protect human, animal or plant life or health as long as those measures are not inconsistent with the Agreement.[64] In particular, they must only be applied to the extent necessary to achieve these objectives and can only be maintained if they are supported by sufficient scientific evidence.[65] Echoing the sentiments of the GATT Agreement, it is also a basic requirement that any measures adopted should not, arbitrarily or unjustifiably, discriminate between Members where identical or similar conditions prevail, or be applied in a manner that constitutes a disguised restriction on trade.[66] In order to promote harmonisation across the globe, the Agreement provides that national sanitary and phytosanitary measures should be based upon international standards, guidelines or recommendations.[67] Three international standard-setting bodies are specifically recognised for this purpose:[68]

The Codex Alimentarius Commission: in relation to international standards, guidelines and recommendations on food safety   The Codex Alimentarius Commission was established in 1963 by the United Nations Food and Agriculture Organisation and the World Health Organisation.[69] It seeks to establish food safety standards in order to protect human health and to ensure fairer competition in international trade. The standards that it sets include standards on issues such as food additives and pesticide residue levels.

The World Organisation for Animal Health (formerly known as the International Office of Epizootics): in relation to international standards, guidelines and recommendations on animal health and zoonoses[70]   The International Office of Epizootics was established in 1924 in order to implement international measures to protect animal health. Since 2003 it has been known as the World Organisation for Animal Health. It currently has 172 member countries.

---

64   The SPS Agreement, Article 2.1.

65   *Ibid.* Article 2.2.

66   *Ibid.* Article 2.3.

67   *Ibid.* Article 3.1.

68   *Ibid,* Annex A, para. 3.

69   For further details on the operation of the Codex Alimentarius Commission, see http://www.codexalimentarius.net/ and also World Health Organisation and the Food and Agriculture Organisation of the United Nations (2005), *Understanding the Codex Alimentarius*, (World Health Organisation and Food and Agriculture Organisation of the United Nations).

70   The World Organisation for Animal Health is usually referred to as the OIE, based upon the French title of its former name, the International Office of Epizootics. For further information about the World Organisation for Animal Health, see http://www.oie.int.

The Secretariat of the International Plant Protection Convention: in relation to international standards, guidelines and recommendations on plant health.[71]　　The International Plant Protection Convention was signed in 1951 and, subsequently, revised in 1979 and 1997. Some 166 countries are parties to it. The Convention seeks to ensure that effective international action can be taken to prevent the spread and introduction of pests of plants and plant products and to ensure that appropriate measures are adopted for their control. The scope of the Convention incorporates both cultivated plants and also natural flora and plant products.

National measures that conform to the standards, guidelines or recommendations established by these bodies are presumed to be consistent with both the terms of the SPS Agreement and with the 1994 GATT Agreement.[72] Article 3.3 of the SPS Agreement also authorises members to introduce measures that provide a higher level of protection than that which would be achieved through applying relevant international standards.[73] However, there must be scientific justification for such measures.[74] Additionally, they must be based on a risk assessment.[75] This risk assessment must take into account a number of factors: available scientific evidence; relevant processes and production methods; relevant inspection, sampling and testing methods; the prevalence of specific diseases or pests; the existence of pest or disease free areas; relevant ecological and environmental conditions; quarantine or other treatment.[76] It must also take into account the following economic factors – potential damage to production or sales caused by the entry; establishment or spread of a pest or disease; the costs of control or eradication within the importing state; the relative cost effectiveness of alternative approaches to risk limitation.[77] As to human health, a WTO Appellate Body has confirmed that the risk assessment should evaluate the 'actual potential for adverse effects on human health in the real world where people live and work and die' as opposed to being laboratory based.[78] WTO panels will consider the adequacy of the risk assessment steps adopted by members when considering whether national measures are compatible with Article 3.3. In *EC-Biotech Products.* The WTO panel also considered the validity of national safeguard laws introduced by a number of European Community Member States to prevent the sale or use of GM products that had already been approved for use by the European Community.[79]

71　For further information on the work of the Secretariat of the International Plant Protection Convention see http://www.ippc.int/.

72　The SPS Agreement, Article 3.2.

73　*Ibid*, Article 3.3.

74　*Ibid.*

75　*Ibid.* Article 5.1.

76　*Ibid*, Article 5.2.

77　*Ibid.* Article 5.3.

78　WT/DS26/AB/R and WT/DS48/AB/R *EC Measures Concerning Meat and Meat Products (Hormones)*, para. 187.

79　See footnote 61 above. The national laws in question had been introduced by Austria, France, Germany Italy and Luxembourg.

In each case these national laws were found to have been based upon inadequate risk assessments and, consequently, to be in breach of Article 3.3.[80] Any national measures adopted under Article 3.3 must also minimise impact upon trade.[81]

Article 3.3, clearly, requires that national measures be scientifically justified and based upon an adequate risk assessment. This was further clarified by WTO panel proceedings brought against the European Community by the United States and Canada, in relation to EU legislation prohibiting the importation of meat or meat products from animals treated with particular growth hormones.[82] The applicants argued that the legislation contravened Article 3.3, as they were neither based upon international standards nor upon any risk assessment.[83] The European Community countered that they were based upon the 'precautionary principle', which they claimed to be a principle of customary international law. Neither the WTO panel nor the Appellate Body addressed the legal status of the 'precautionary principle'. However, both found the EU measures contravened the SPS Agreement. The Appellate Body accepted that the SPS Agreement reflected the 'precautionary principle', but found that this did not enable members to use the principle to justify measures that were, otherwise, inconsistent with the Agreement.[84] Consequently, the 'precautionary principle' could not override members' obligations, under the Agreement, to ensure that sanitary and phytosanitary measures were based upon an appropriate assessment of risks to human, animal or plant life or health.[85]

The 'precautionary principle' is certainly evident in the fact that the Agreement does authorise countries to adopt provisional sanitary or phytosanitary measures when insufficient scientific evidence is available.[86] These provisional measures must be based upon relevant information that is available both nationally and internationally.[87] Countries adopting such provisional measures are also placed under an obligation to seek the additional information necessary to enable them to make a more objective assessment of the risk poses and to review their provisional measures within a reasonable time frame.[88]

Equally, the SPS Agreement seeks to protect international trade by promoting the 'principle of equivalence'. Countries are required to accept sanitary or

---

80   *Ibid.* paras. 7.3076, 7.3096, 7.3110, 7.3117–7.3119 and 7.3146–7.3150. For an analysis of the panel decision on these provisions see Cheyne, I., footnote 61 above, 60.

81   *Ibid.* Article 5.4.

82   As per footnote reference 78 above.

83   The legislation in issue was Directive 81/602 EEC of 31 July 1981, Directive 88/146/EEC of 7 March 1988, Directive 88/299/EEC of 17 May 1988 and Directive 96/22/EC of 29 April 1996 prohibiting the administration to farm animals of substances having a hormonal action and of substances having a thyrostatic action (which repealed and replaced the previous measures).

84   Footnote 78 above, para. 93.

85   *Ibid.*

86   *Ibid.* Article 5.7.

87   *Ibid.*

88   *Ibid.*

phytosanitary measures adopted by other nations as equivalent, even if they are different from their own or those used by other countries, if the exporting nation can objectively demonstrate to the importing nation that its measures achieve the same standards of protection.[89]

The Agreement on Technical Barriers to Trade

The TBT Agreement also enables countries to introduce measures to protect human, animal or plant life or health and the environment, so long as they do not constitute a means of arbitrary or unjustifiable discrimination between countries or a disguised restriction on international trade. The focus in this Agreement is upon technical regulations and standards concerned with product characteristics and their processing and production methods. These will include product standards or processing and production requirements relating to agricultural produce. Under the Agreement a technical regulation is defined as being:[90]

> a document which lays down product characteristics or their related processes and production methods, including the applicable administrative provisions, with which compliance is mandatory. It shall also include or deal exclusively with terminology, symbols, packaging, marking or labelling requirements as they apply to a product, processor or production methods.

Countries must not apply technical regulations to imports that are less favourable than those applied to products produced nationally.[91] Additionally, whilst the protection of human health or safety, animal or plant life or health, or the environment are recognised amongst the list of legitimate objectives for which they can be adopted, technical regulations must not be more trade restrictive than necessary in order to fulfil legitimate objectives.[92] The technical regulations must be based upon a risk assessment that is, in turn, based upon available scientific and technical information, related processing technology and the intended end use of products.[93]

The Agreement requires countries to ensure that technical regulations are not maintained longer than required or if their objectives can, subsequently, be attained by less trade restrictive methods.[94] As in the case of the SPS Agreement, the TBT Agreement also seeks to harmonise technical standards by requiring countries to base their technical regulations, as much as possible, upon any relevant

---

89   *Ibid.* Article 4.
90   The TBT Agreement, Annex I.
91   *Ibid,* Article 2.1.
92   *Ibid.* Article 2.2.
93   *Ibid.*
94   *Ibid.* Article 2.3.

international standards that either exist or whose completion is imminent.[95] Mirroring the SPS Agreement, where a technical regulation is introduced to protect a legitimate objective and it accords with international standards, it shall, rebuttably, be presumed not to create an unnecessary obstacle to trade.[96] Countries are also required to play their part in harmonising technical standards by assisting appropriate international standard setting bodies.[97] Unlike the SPS Agreement, particular bodies are not identified within the TBT Agreement, but for agricultural produce the Codex Alimentarius Commission will, once again, have an important role.

Countries are not always bound to follow international standards in developing their technical regulations. The TBT Agreement provides that no obligation arises in situations where the existing international standards would be an ineffective or inappropriate way to achieve a legitimate objective.[98] By way of example, it provides that this might be because of fundamental climatic or geographical factors or fundamental technological problems.[99] However, any country preparing, adopting or applying a technical regulation, that may have a significant effect upon trade with other countries, can be required, by another member country, to explain the justification for that measure.[100]

Again, as in the case of the SPS Agreement, countries must accept the technical regulations of other countries as being equivalent to their own, where they are satisfied that these regulations fulfil the same objectives as their own.[101]

The Agreement on Trade Related Aspects of Intellectual Property Rights

The TRIPS Agreement, again, stems from a desire to protect legitimate objectives, in this case the protection of intellectual property rights, whilst at the same time minimising their impact upon international trade. The TRIPS Agreement covers a wide range of intellectual property rights – copyrights, trademarks, industrial designs, patents and so forth. However, the most important aspect of the Agreement, for agriculture, is the fact that it also deals with laws governing the use of geographical indicators. These are defined as being[102]

> …indications which identify a good as originating in the territory of a [country], or a region or locality in the territory of a [country], or a region or locality in that

---

95   *Ibid*, Article 2.4.
96   *Ibid.* Article 2.5.
97   *Ibid.* Article 2.6.
98   *Ibid.* Article 2.4.
99   *Ibid.*
100   *Ibid.* Article 2.7.
101   *Ibid.*
102   TRIPS Agreement, Article 22.1.

territory, where a given quality, reputation or other characteristic of the good is essentially attributable to its geographical origin.

The TRIPS Agreement sets out a range of provisions designed to govern national laws concerning the use of geographical indicators. For example, countries must ensure that their laws enable interested parties to take action to prevent methods from being used in the designation or preparation of goods that would mislead the public as to their geographical origin, by indicating or suggesting that they originate from a geographical area other than their true place of origin.[103] Additionally, the TRIPS Agreement provides for countries to refuse, or annul, the registration of trademarks containing or consisting of a geographical indication for goods, where the goods concerned do not originate in the territory indicated and the use of the indicator is likely to mislead the public as to their true origin.[104] Equally, the TRIPS Agreement established the basic rule that national intellectual property laws should provide nationals of other countries a level of treatment that was no less favourable than that which was accorded to their own nationals.[105] In 2005, a WTO Panel found that the European Community's Regulation 2081/1992,[106] on the protection of geographical indicators and designation of origin for agricultural products and foodstuffs, failed to adhere to this aspect of the TRIPS Agreement.[107] The panel found the Regulation to be discriminatory as applicants from third countries could only register geographical indications within the European Community if their governments adopted systems to protect geographical indicators that were equivalent to that adopted by the European Community. Even then, these applications had to be made through their governments. Consequently, foreign nationals were found not to have the same level of guaranteed access under the Regulation as European Community nationals. This finding resulted in the Regulation being repealed and replaced by Regulation 510/2006, which now provides equal access to applicants from countries outside the European Community.[108]

---

103   *Ibid*, Article 22.2.

104   *Ibid*. Article 22.3. Under Articles 24(5) this does not apply to trademarks registered in good faith before the TRIPS Agreement applied in the country of registration or before the geographical indication was protected in the country of origin. Additionally, Article 24(6) provides that a member country has no obligation to apply this provision where geographical indicators relating to another country are used for goods and services and are identical to terms that are used in common language as the common name for those good or services in that, former, country.

105   Article 3.

106   [1992] OJ L208/1.

107   WT/DS174 and WT/DS290: *European Communities – Trademarks and Geographical Indications* (15 March 2005). The panel decision was based upon applications lodged by the United States and Australia.

108   [2006] OJ L93/12.

Importing Agricultural Produce into the European Community

The European Community operates a range of controls designed to prevent animals or plants carrying infectious diseases from being imported into the Community. The measures that apply to the importation of live animals can be grouped into those concerning aquaculture,[109] cattle,[110] horses,[111] sheep and goats,[112] pigs,[113] pets,[114] poultry and eggs[115] and other livestock.[116] Similarly, animal health rules also apply to the import of animal products that are intended for human consumption.[117] The following section describes the measures in place for importation of live cattle and of meat products.[118] These measures are based around three main controls – the identification of specific third countries from whom exports may be made to the European Community, the use of veterinary health certificates accompanying the livestock or meat products and a veterinary inspection at an European Community border post prior to entry into the Community.

---

109   Council Directive 91/67/EC concerning the placing on the market of aquaculture animals and products, [1991] OJ L46/1.

110   Council Directive 2004/68 laying down animal health rules for the importation into and transit through the EU of certain live ungulate [hoofed] animals, [2004] OJ L139/321.

111   Council Directive 90/426 on animal health conditions governing the movement and import from third countries of equidane, [1990] OJ L224/42.

112   Directive 91/68/EC on animal health conditions governing intra EU trade in and imports from third countries of ovine and caprice animals, [1991] OJ L46/19.

113   Directive 72/462/EEC on health and veterinary inspection problems upon importation of bovine, ovine and caprice animals and swine, fresh meat or meat products from third countries, [1972] OJ L302/28, as amended.

114   Regulation 998/2003/EC on animal health requirements applicable to the non commercial movement of pet animals, [2003] OJ L146/1.

115   Directive 90/539/EEC on animal health conditions governing intra EU trade in and imports from third countries of poultry and hatching eggs, [1990] OJ L303/6.

116   Directive 92/65/EC laying down animal health requirements governing trade in and imports into the EC of animals, semen and embryos not subject to animal health requirements laid down in specific EC rules, [1992] OJ L268/54.

117   Directive 2002/99/EC laying down animal health rules governing the production, processing, distribution and introduction of products of animal origin for human consumption, [2002] OJ L18/11. For milk and milk products additional requirements, such as those concerning treatments used in the manufacture of milk and milk products are set out in Commission Decision 95/340/EC, [1995] L200/38.

118   See also European Commission, 2007. *General Guidance on EU Import and Transit Rules for Live Animals and Animal Products From 3rd Countries*, Brussels: European Commission, available at http://ec.europa.eu/food/international/trade/guide_ thirdcountries2006_en.pdf [last accessed 10 August 2009].

Recognition of third countries entitled to export to the European Community

Livestock and livestock products can only be imported into the European Community if they come from a third country that is authorised to export those animals or produce to the Community. The European Community maintains lists of countries, or regions of particular countries, from which exports can be made, for each livestock category and for animal products. Inclusion upon these lists is based upon a European Community audit that seeks to ensure that adequate animal health measures are in place there. This audit will consider factors such as:[119]

- The legislation, in place in that country, on animal health and welfare
- The organisation and powers of its veterinary authority and the degree of supervision that it is subject to
- The animal health requirements that apply to the production, manufacture, handling, storage and dispatch of produce intended for the European Community
- The assurances that the competent veterinary authority can give concerning compliance withanimal health conditions
- Experience previously gained in importing livestock or animal produce from that country
- The results of EU inspections and/or audits carried out in that country
- The health status of livestock, other than domestic animals and wild life in that country
- The regularity, speed and accuracy with which that country supplies information concerning the existence of infectious or contagious animal diseases on its territory
- The rules concerning the prevention and control of infectious or contagious diseases that are in force in that country

In addition the European Community may also require compliance with further, specific conditions, on animal health issues. These could require, for example, the necessity that the exporting country be free of particular diseases, such foot and mouth disease, blue tongue or avian flu.

Veterinary Health Certificates

In addition to the approval of the exporting country, each export must also be accompanied by a veterinary health certificate, completed by an authorised

---

119 See, for example. Directive 2002/99/EC laying down the animal health rules governing the production, processing, distribution and introduction of products of animal origin for human consumption, [2003] OJ L18/11, Article 8 or Directive 2004/68/EC laying down animal health rules for the importation into and transit through the EU of certain ungulate animals, [2004] OJ L139/321, Article 4.

veterinary official in that country. For example, in the case of exports of live cattle,[120] this will certify that the cattle come from a territory free of specific diseases and in which vaccination has not been carried out,[121] that the animals themselves were inspected prior to movement and were healthy and that the conditions in which they were transported meet animal welfare standards set out in Directive 91/628/EC.[122]

Border Inspections

As a third barrier against infectious or contagious diseases that might affect livestock or humans, no consignment of livestock or animal produce can be introduced into the European Community unless it has been subject to a veterinary inspection at a Border Inspection Post.[123] At this border post the relevant Member State's veterinary officers will be required to ensure that the livestock or livestock produce is healthy and complies with European Community legislation.

Directive 2000/29/EC on protective measures against the introduction, into the European Community, of organisms harmful to plants or plant products and against their spread within the European Community (the 'Plant Health Directive') sets out measures that apply to both imports from non-Member States and also intra-Community trade.[124] The Directive was adopted to consolidate European Community law in this area.[125]

The Plant Health Directive identifies particular harmful pests that must not be allowed to enter or circulate within the European Community and also those that must, similarly, be banned if found to be present in particular plants or plant products.[126] It also provides a list of plants and plant products that may not be imported into the EU from particular third countries.[127] Additionally, the Plant Health Directive also identifies requirements that must be met by specific plants or

---

120   See Regulation 2004/68/EC laying down animal health requirements for the importation into and transit through the EU of certain ungulate animals, [2004] OJ L139/321, Article 11.

121   *Ibid.* Annex II, lists foot and mouth disease, vesicular stomatitis, swine vesicular disease, rinderpest, peste des petits ruminants, contagious bovine pleuropneumonia, lumpy skin disease, rift valley fever, blue tongue, sheep\pox and goat pox, African swine fever and classical swine fever.

122   [1991] OJ L340/17.

123   As to the detailed requirements for these veterinary inspections see Directive 97/78 laying down the principles governing the organisation of veterinary checks on products entering the EU from third countries, [1997] OJ L241/9.

124   Council Directive 2000/29, OJ [2000] L169/1.

125   This had previously been set out in Council Directive 77/93/EEC on protective measures against the introduction into the EU of organisms harmful to plants or plant products and against their spread within the EU, OJ [1977] L26/20, as amended.

126   Council Directive 2000/29, annexes I and II respectively.

127   *Ibid.* Annex III.

plant products before they can be allowed into the EU or allowed to cross internal frontiers and specifies those for which special arrangements may be made.[128]

The Plant Health Directive implements these controls through a system of inspections operated by officially designated plant protection services in each Member State. It identifies the plants, plant products and other objects (such as packaging materials) that must undergo a plant health inspection before being allowed to enter the European Community.[129] Member States must subject these items to meticulous inspection before they can enter the European Community.[130] In addition, the items must be accompanied by a phytosanitary certificate, made out not more than 14 days before the date on which they left the exporting country, which confirms that they comply with the European Community's plant health law. [131] Subject to a number of minor derogations, these provisions will apply to all plants, plant products and other objects identified by the Plant Health Directive.[132] If the inspection confirms that the plant materials concerned do comply with European Community's Plant Health Law, then a 'plant health passport' will be issued by the designated plant protection service of the importing Member State.[133] This document will be evidence of the fact that the material does comply with European Community Plant Health law and, with one exception, enables it to move freely throughout the European Community. The only exception relates to the fact that the Plant Health Directive also enables the European Community to recognise protected zones in which even more stringent measures are applied. For example, the United Kingdom and Ireland are each recognised as protected zones in relation to the spread of the Colorado beetle, which poses a threat to potato crops.[134] Protected zones are defined as being zones within the European Community within which either:[135]

a. …one or more harmful organisms referred to in [the] Directive, which are established in one or more parts of the [European Community], are not endemic or established despite favourable conditions for them to establish themselves there, or

---

128   *Ibid.* Annexes IV and VI respectively.

129   *Ibid.* Annex V.

130   *Ibid.* Article 13(1).

131   *Ibid.* The article requires that the certificate be based upon the model set out in the annex to the International Plant Protection Convention.

132   According to Article 13 (4) to (6) derogations may apply to shipments of plant materials between two EC territories via a non Member State, to small quantities of material intended for non industrial/non commercial use or consumption during transit and to materials to be used for trial or scientific purposes. Such derogations can be applied when no risk exists of the spread of harmful organisms.

133   Directive 2000/29, Article 10.

134   See Black, R., (2003), The Legal Basis for Control of Imports of Animals and Plant Material into the United Kingdom, [2003] 5 *Environmental Law Review* 179 at 186.

135   Directive 2000/29, Article 2(1).

b. there is a danger that certain harmful organisms will establish, given propitious ecological conditions, for particular crops, despite the fact that these organisms are not endemic or established in the [European Community]

The Plant Health Directive identifies the protected zones recognised by the European Community and sets out the additional requirements that apply, in relation to the exclusion of particular pests, within each zone.[136] Plant materials originating in third countries will be inspected for compliance with these additional requirements, whilst materials in circulation within the European Community will only be able to enter the protected zone if accompanied by a plant passport valid for that zone.[137]

---

136 *Ibid,* Annexes I–IV.
137 *Ibid.* Articles 10(2) and 13(8).

Index